Table of Contents 12/26/17
Introduction 3
Background 4
Beyond Epigenetics 8
The Gene Code/ the Jeans We Wear 9
Genetics in Medicine (GIM) 12
Epigentics and Methylation Therapies (EMT) 14
Dietary Oils 14
Saturated Fats 26
Cancer Cures 28
Common Cold and Flu 36
MS 36
Diagnosis 37
Quantum Healing (QH) 40
Fractals 41
Vibratory Frequencies 48
Dis-Ease 49
Cellular Energy Systems 51
Bio Subsystems (QB) 52
Diet 53
Telomeres 55
Packaged Supps/ Heart 56
Energy Cycles 61
Living Water 63
MCA 65
Disease Management 70
HBP 72
EM Devices 73
Balance 75
Histamines 75
MSM Trial 83
Parasitic Diseases 87
MSM Protocol 91
Epigenetics/ Diagnosis 94
Quantum Healing QH Aspects 95
Plant Growth and Vibrations QB 105
What is Coming? 108
In Closing 111
Videography/ Bibliography 112-3

About the Author

James Robert Clark, is a close friend of Dr. Richard Price whom he has worked with on various quantum fractal projects (but no direct business association). He and Richard together developed many of the working models and final products discussed in this book.

Has written five other books as listed in the bibliography that teach Spirituality, Healing, Health, and Longevity.

Holds a BAS in art and a 1st professional degree in architecture and was first registered in 1983. Is a practicing registered professional architect with projects in several states, but mainly Delaware, Maryland, and Virginia. Is an inventor with one US patent in his name and several contributions.

Interests and hobbies include, Painting, Figure Drawing, Ceramic Sculpture, Woodworking, Furniture Design, Sailing, Tennis, Flying, Boatbuilding, Fishing, Skiing, Sports Car Restoration, and Gardening.

Dedication

This Book is dedicated to Dr. Richard Price, PhD, himself. As a dedicated and conscientious healer and friend of mankind.

I miss his wife, Joanne, to whom I dedicated my first book. She died of associated back problem treatments during the writing of the initial book, "It's the Liver Stupid."

I wish my good friend Rich well.

Between the Jeans
Epigenetics, Awareness, Antiaging, and Health
Exceeding Current Science

Introduction:

Welcome to "Between the Jeans," a complete rewrite of "Beyond Epigentics," which was the third in this original series of books that centered on high level MSM. Since that edition, much has transpired and this book incorporates the most profound advancements in healing known to man, maybe ever, but certainly, in modern history. This new rewrite and update of the original includes completely new and original technology not contemplated during the first book writing.

Herein, we again revisit this perceived fractally-based quantum science as taught by Dr. Dan Nelson, PhD. Then, the work of Dr. Bruce Lipton, PhD is combined with the work of my friend, Dr. Rich Price, PhD, whose spiritual ability can rise far beyond the subconscious level and diagnose illness from the higher planes of what some term as God-Consciousness.

Finally: To get a full understanding of this book, you must watch or have watched a few of my reference videos that include some thirty of my videos on You Tube, thus witness some of what I describe here. To access them, search James Robert Clark Quantum Science.

Background:

The first two books were not totally original ideas in our search for health. Still, High Level MSM, as introduced, was virtually unknown to nearly everyone and certainly was unknown to even the most informed professionals when my first book came out.

However, the Kentucky horse farmer who introduced it to me has proven to have been thoroughly on top of his game and, as new facts are revealed with new studies and reader feedback, his wisdom was indeed far reaching, original in its detail, and, up until now, has proven itself totally accurate, but feedback proves it to be well beyond what he reported and, at this point, it is hard to find a disease or condition that it does not positively affect, from heart disease to prostrate problems, people just improve. So, my first book, "It's the Liver Stupid" explained and introduced this HL(high level) MSM protocol. It is further expanded upon herein with many new supplements, work, studies, and commentary. It always included Undenatured Whey Protein Isolates, and the Low Carb Diet protocols along with many adjutants, but now adds "Living Water" here and in the 5th edition. Living Water as reported herein alters plant and animal biology in ways previously unknown to modern science through Quantum Biology (QB).

My second book, "Methylation, Awareness, and You" explained the cutting edge methylation sequences, QB, and epigenetics, thus revealed the mechanisms behind, especially, the HL MSM Protocol of the first book.

Rich first learned about Quantum Healing (QH) after attending a conference in Niagra Falls in 2013 after this author encouraged him to do so. Dan Nelson is probably the foremost quantum scientist on earth in terms of the workings of cutting edge fractal/quantum technology theory. However, the actual work that Rich is doing now, no doubt now surpasses Dan's in application. Rich developed his same CP4U-F (Crystal Pack 4 Unit Fractal device) for cars, but with different proprietary combinations (to match conditions) for human and plant health. That is, he selects the conditions to contain the information required to create the vibrations. We used four 8" lab crystals in the car program. Hence the name. The CP4U was adequate for 2 to 5 Liter cars. The point here is that size matters when it comes to crystals since they are intelligent storage devices, essentially hard drives. Think of lab crystals as large but intelligent microchips that actually adapt to conditions.

So we first used fractals as machine enhancement devices that improved car performance. They, in a true sense, "cured" endemic engine ills. These systems proved very predictable and reliable and even repaired badly running older cars to some degree. In every case, they caused our test VW Jetta TDI's to record mileage figures never previously reported or even dreamed of by mainstream science, no matter what the driving conditions. When you look at my videos (James Robert Clark) on You Tube, you often see figures well into the 200 mpg range even under light acceleration and trip mileage under controlled sustained tests showed 100mpg at 80mph in one dramatic two day test. This is in your face proof, not idle claims.

Next we moved this technology from healing cars to people and then to plants to make it even more profound. Herein, as a result, we discuss this perceived new, fractally-based quantum science as taught by

Dan Nelson with learned and observed results combined with Rich's intuitive ability... ground breaking science.

This book, "Between the Jeans" (Beyond Epigentics II?) then discusses this original, fractally-based method of diagnosis as devised by my friend, Dr. Richard S. Price, PhD to whom this book is dedicated. Using this, he combined his fractal healing system, loosely based on the work of Dan Nelson and expanded on it in a way that no one else likely could. Further, Rich has since taught his technology to others in classes, so it is teachable to a select few as Dan's is.

With this, I submit that this new work takes healing, wellness, and thriving to realms heretofore unattainable by conventional science and expands on it using Quantum Science, Epigentics, and more.

While Epigenetics is still very much cutting edge and best expressed by Dr. Bruce Lipton PhD, the work herein was never documented prior to this book. Furthermore, Rich has applied it to hundreds of people with such parasitic diseases as Lyme and Malaria and their related infections and has successfully resolved them, sometimes in just a matter of minutes or hours. Finally, I have personally witnessed some of these events real-time.

With this, I end this book with a report on the incredible 400% plant growth results obtained with water exposed to these fractals and fractalized crystals as reported. With plants, the results are more easily documented. While still variable in their outcomes, plants are totally within their own control and the results are in your face. Here, no outside subjective error can easily be argued as opposed to intervening with cars and people.

But, no matter what the target: a diesel engine, a human body, or a strawberry plant, these fractals (QH) work extraordinarily well and at levels previously unknown or documented by anyone as far as we know.

Beyond Epigenetics

The above introduction begins to explain how Dr. Rich Price PhD, introduced above, can actually rise above the subconscious mind, Lipton's territory, and make actual changes to it so as to affect mental and physical health in ways previously never understood by science. While this ability can be and is taught to a select few, it is not likely to ever be within the total knowledge of most health care practitioners. However, the goal herein, is to help everyone understand these levels of healing and to move this incredible science forward.

Moving the Bar/ Visualization

On the opposite end, the popular concept that someone can will themselves well is equally incorrect. No one can make subconscious changes intentionally (consciously). What we can do is to set up the conditions to make them occur as Lipton tells us. This is creative visualization, akin to creative daydreaming. Our spiritual abilities on the subconscious level and above are incredible and now are observable in talks and demonstrations by Dr. Bruce Lipton and others on You Tube videos.

The Gene Code/ the Jeans We Wear
http://en.wikipedia.org/wiki/Genetic_code

The genetic code is the set of rules by which information is encoded within genetic material, that is, DNA and mitochondrial RNA (mRNA) sequences are translated into proteins by living cells. Biological decoding is accomplished by ribosomes and these link amino acids in an order specified by mRNA, using transfer RNA (tRNA) molecules to carry them and to read the mRNA, three nucleotides at a time.

It is important to note here that the genetic code is known to be very similar among all organisms and can be expressed in a simple table with 64 entries (per Wikipedia). That is, all animals wear the same style of Jeans. Furthermore, this code, as applied to the human genome, is far more simple than previously anticipated prior to gene sequencing. In fact, in sequencing the code, it was discovered that the human genome was no more complex than that of far more simple animals. Interestingly, in some cases, our code is less complex than some worms to the surprise of virtually all scientists involved prior to the sequencing.

Ours are not designer Jeans. That is, the anticipation was that it would be many times that of any other biology on earth. The discovery reported herein, then, is that it is "quantum events," or epigenetics that actually guide our complexity far more than genes. Thus, we are in control of our demise in ways that were never before anticipated as many cutting edge scientists have since discovered and demonstrated.

Furthermore, prior to sequencing, the great hope in decoding the sequences was that this would eventually result in curing all disease, especially autoimmune diseases and cancer, that they could restyle them into designer jeans. None of this occurred, of course, but as the below publication predicts, today's science still grasps this "great hope" tightly even though it simply can never be. So we see that science today has become more of a wish and a dream than a fact as we waste billions of dollars on hopes that show no promise as Dr. Bruce Lipton and those who have followed his footsteps point out so cleverly in their talks.

In the field of cancer, my greatest hope of resolving simply, very little has changed following the sequencing. The "new" chemo treatments, are now called "targeted gene therapies" (or very high styles of Jeans). **http://www.cancer.gov/cancertopics/treatment/types/targeted-therapies/targeted-therapies-fact-sheet** are still basically poisons to the human body just as was the first chemo treatment, "mustard gas" of WWI. To quote the above link: "***Scientists had expected that targeted cancer therapies would be less toxic than traditional chemotherapy drugs because cancer cells are more dependent on the targets than are normal cells. However, targeted cancer therapies can have substantial side effects***." The point here is that they all still carry these side effects just as their own literature points out and they are life threatening when taken as a protocol... again, the same old Jeans warmed over.

Should you disagree with the above, read the "side effects" of these individual chemodrugs, as published by each of their pharmaceutical manufacturers. The fact remains that these cancer drugs are the same killers as ever. So the bottom line is: Most all chemodrugs will often initially cause tumors to temporarily recede just as mustard gas did years ago. Eventually, when taken long-term as virtually all of these drug companies report of their own drugs, the negatives appear. Furthermore, they can and generally will cause death, given the opportunity.

So at this point, the idea that genetics makes cancer drugs more safe has now generally become a marketing ploy and not the established fact that they want you to believe. Moreover, these, so called, targeted drugs are among the most expensive in the world and anything to help sell them will be touted by their owners. So, no, this is not the scientific advancement that the public has been led to believe. This is just an over anticipated scientific advancement that never panned out... an idea and hope that was generally anticipated by the scientific world.

Genetics in Medicine (GIM)

http://www.nature.com/gim/index.html
Genetics in Medicine (GIM) is the official journal of the American College of Medical Genetics and Genomics. This monthly journal teaches a genetic medicine, including topics on chromosome abnormalities, metabolic diseases, single gene disorders, genetic susceptibility to common complex diseases, etc.

When this gene code was broken as described above, there was great hope. This and similar publications still teach that something will materialize. However, despite this huge setback, what has actually occurred is quite useful and as follows, yet none of this was anticipated by the above work (in opposition to the their setback):

Thus, the idea was that the code was determined and everything was set, as their own literature points out, but this was as far as it went. Again, they are still the life threatening killers when taken as a protocol.

Should you doubt the above, read the "side effects" of these individual chemodrugs, as published by each of their pharmaceutical manufacturers. Therefore, they remain the very same killers that they always were by varying degrees.

So, while we will never see the scientific advancement that the public has been led to believe, below is what actually occurred that can give us hope: This is the good that came out of this as we move forward. You can take the following to the bank:

Epigentics and Methylation Therapies (EMT)

One of the huge strides that rose out of this gene sequencing was in our understanding that we, as individuals, can actually change gene sections (known as SNIPS) of the gene code on a subconscious level. This will be expanded on here in various ways, But watch Dr. Bruce Lipton's videos and read his books on this topic.
https://www.youtube.com/watch?v=XjTXzlf4Spw

Dietary Oils... The Good and the Bad

The important aspect of the story below is that it demonstrates just how controlling the commercialized side of health and especially the food and drug industry is and has been historically. Margarine, the topic below, was first introduced well over a hundred years ago and with it ushered in the idea that cholesterol was a heart health problem.

Margarine, the Killer Substitute

The idea was to scare the population into acceptance of this garbage as a food. In fact, it still manages to be sold as a healthy food substitute for butter despite the obvious deception used in selling the product: Margarine is quite simply, whipped-up cottonseed oil now combined with some dairy product to make it more palatable. Cottonseed oil is (wouldn't you know it), an unregulated waste by-product with added inherent toxins on that side also. Not grown as a food (much like tobacco), FDA requirements are vague or nonexistent for cotton crops.

So, even today, cotton production waste has been pawned off as a health food with clever marketing. There is nothing inherent in this cotton seed that could even remotely be considered healthy (or even considered as food), yet it is still being sold and touted today as a heart healthy food and butter substitute that is smiled on by the FDA.

Below is the author's version of this margarine story that is repeated many times on the web in various versions, but edited
https://en.wikipedia.org/wiki/Margarine
http://en.wikipedia.org/wiki/Cottonseed_oil:

The dairy industry was extremely powerful in the late 1800's. Thus, they controlled and affected dairy and marketing practices with an iron hand.

The cotton producers had cottonseed oil by-product, waste, that they were unable to even dump it into the ground. It was an ugly white/gray material with no resemblance to butter and that no one would eat. However, someone got the idea that they could whip it up, color it, and sell it as a butter substitute. This, as a way to get rid of the mess and even profit in the process.

So they came out with new less objectionable appearing products and, along the way, found a PhD scientist whom they managed to get on the cover of Life Magazine, Ancel Keys, who offered it up as a cure for heart disease. His age-old theory derived from a far earlier fat consuming rabbit study was accepted by the AMA and became known as the "Cholesterol Theory."

You can read about the man behind this, Keys and his now infamous theory here: **http://junkscience.com/2013/05/09/putting-a-stake-through-the-cholesterol-theory-of-heart-disease/**

By 1955, the margarine industry was even allowed to color it yellow. So margarine was pushed onto the public as one of the great ways to beat heart disease. Along the way, all unsaturated fats took an enviable marketing position alongside Margarine. So corn, canola, vegetable, peanut oil, and several others became known as the heart-healthy oils and such things as Coconut Oil, mostly saturated were shunned.

Digressing, the butter industry would allow no part of this and they came back with a vengeance. Their first move was to disallow the cotton seed people from coloring their ugly mess with yellow food dyes to make it look like butter. So the answer from the cottonseed people was to provide a coloring "kit." This kit included a packet of their yellow food dye that the housewife mixed in with the cotton seed oil after buying it to make it look like butter.

Thus, they proceeded to sell their margarine colored yellow and today have even come out with such clever ads as. "I can't believe it's not butter," as they continue their charade to push it as heart-healthy. Along with margarine came our deep distaste for such time-honored fats as lard and bacon fat which became known as a heart attack foods and still hold that reputation.

However, please believe this: It is "not butter" and it will eventually help kill you if you use it long enough. In fact, Cholesterol, as discussed herein in several places and touted by the wonderful Dr. Stephanie Seneff, is a major part of your brain (Alzheimer's anyone?). As she points out in her many You Tube videos: Cholesterol is an incredibly essential part of your diet. If it is lacking, you will become diseased. Your mind will deteriorate, your circulatory system will harden, and your heart will fail from overwork, just as she and others have observed and predicted. Also, it offers some protection from cancer and especially skin cancer since it combines with sunlight to produce vitamin D.

Most doctors today can even order the above fats/ killers from best too worst, but they are our problem along with unsaturated transfats. No one then even paid attention to the transfats that they become at high temperatures when they were used as cooking oils, but cooks loved the fact that they seldom became rancid (spoiled) and that they could fry with them for weeks, so they became the standard for deep fryers in all restaurants and still are today.

Now, of course, they have managed to convince the entire medical industry (with the advice of Keys) that saturated fats and cholesterol are the killer and even separated it into two groups (HDL/ LDL), the well

known good and bad cholesterol.
http://healthimpactnews.com/2013/there-is-only-one-type-of-cholesterol-all-of-it-beneficial/

Also, read what Wikipedia is now telling us as the word gets out on fats:
https://en.wikipedia.org/wiki/Saturated_fa

Listen to Dr. Stephanie Seneff's interview https://www.youtube.com/watch?v=I-jJn-4jUxg Above, she discusses the need for higher cholesterol (and as much as you can get). In fact, including more cholesterol in your diet, as she discusses, will result in the lowering of heart disease rates and blood pressures. Of course, this is just the opposite of what we are taught by most mainstream doctors today, but her stand will eventually become a very well established fact as we move forward. Note that all animals will eat the fat, bones, and organ meats first that you throw out, just as you pass up the fats and eat the lean cooked and even burnt meat from your grill. So I ask, who is the more intelligent?

The best saturated fat (8-13% monosaturate), however, may just be coconut oil and it seems to make many things taste better when used in cooking. Additionally, it is very resistant to high heat.

Thus, it does not become a transfat easily. Therefore, it is often recommended today by informed healthcare professionals, even though many med schools and dieticians still will shun it as they spread their dangerous dietary information after a half century of bad dietary and medical advice.

Today, the AMA and FDA, generally, still back this unsaturated fat story, along with Statins, of course, the most successful drugs ever sold)**:**
http://www.cottonseedoiltour.com/facts/
http://www.cottonseed.com/publications/facts.asp

A longer and more involved historical version of the above story is available at: http://mentalfloss.com/article/25638/surprisingly-interesting-history-margarine
as Dr. Mercola's dialog agrees with the above commentary at:
http://articles.mercola.com/herbal-oils/cottonseed-oil.aspx

Cancer Theory Revisited

This above story was introduced early-on as an obvious scam in hopes that the eyes of the readers with doubts will be opened. That is, you will read the references and thus will begin to understand just how badly we have been duped by the Food and Drug Industry, AMA and FDA. These lies carry multi-billion dollar price tags, so they will remain difficult to be unraveled, but this is the reality and the truth that must eventually come out.

The premise herein is that all diseases are curable, but that cancer itself is not really a disease since it is, in fact, toxin based. If this is the case and the parasites (the disease aspect) move in afterwards, it is purely an environmental condition like radiation poisoning in cause. More on this as we move along. The above transfat story, a key element in many diseases, was the first bolt from the mainstream teaching, but more will follow as we go into this possibly more horrific lie with cancer as follows:

Cancer and Fats

Now realize that cancer, to begin with, is multifaceted as a disease. Today, this so-called disease, cancer, is seen as rapidly increasing and uncontrolled cell division with no real known cause other than bad genes and, possibly, too many toxins. So, no one today has a clue as to why cancer occurs for the most part. Seldom does anyone even offer a reason other than cigarettes or asbestos. Instead, what you hear is that it is a very complex disease that stems mostly from genes having gone awry, even though many billions of dollars have been spent in studying it (and certainly more than on any other disease).

Below, I offer the cause, as I look at it in a very different light, based on the idea that our bodies are actually quantum machines that generally heal themselves if given the opportunity, as contended herein.

Agreed, this theory is a radical departure from mainstream AMA and FDA descriptions, but it does make sense compared to the others who throw up their hands when are asked to analyze the disease and begin discussing why the root cause rather than observations (effects):

Thus my stated cause: **Cancer is a naturally occurring reaction to toxic chemical build-ups and environmental poisons (generally combined with a compromised immune system)**. It has always been around to some degree because toxins do occur in nature, but it has increased at alarming rates since the industrial age as our toxins have increased. Furthermore, due to the fact that the condition always lowers immune response, it is accompanied by invasive nanobacteria, fungal, viral infections, and parasites that pile onto the unprotected organism (as they always will). Today, the curve is moving up alarmingly, yet reports play this increase down, saying that we are winning this war. So what is changing? For one thing, the rates of toxic exposure likely have a hand in where cancer sites occur, but there are obviously no statistics on this to correlate.

As to healing cancer, when the co-infections are removed, it gives the body a far better chance of overcoming the offending toxins that initialized the problem, as all organisms are designed to do. In fact, as Royal Rife proved in the 1930's: In most cases, the people relieved of opportunistic invaders survive and the cancer disappears.

Thus, I submit that the cancer will generally dissipate if either the invaders or the toxins are removed as long as the diet is sufficient to provide the dietary sulfur, vitamins, minerals, and fats needed for overall wellness and a stronger immune system. Therefore, in every case, if a person is relieved of one or all conditions and it has the required nutrients, there will be no more, so called "disease."

The problem here is (and it has been true since ancient Greek reports), that cancer has been considered a biological condition, but more recently, a genetic mistake.

However, given the above, my theory holds that cancer is just another system of biological survival. The bottom line then, delving deeper, is that cancer is a condition in which the body is attempting to relieve itself of toxins from an overloaded liver. In this process, these host invaders infect the body as secondary infections (that in the 1930's, Rife interpreted these as primary, in his work). Your body will do anything it can to survive just another minute and, by this theory, cancer is a survival mechanism.

Below are some rather well established and common generalities pointed out to back up the above proposed theory:

1. There is no such thing as cancer of the heart. Why? In its infinite wisdom, the body knows that the heart is essential, so it protects it at all cost. If the site were not discretionary, how could the body make this determination? The heart circulates blood throughout the body. All cancer specialists agree that toxins play a role in cancer just as it did historically with the nuclear radiation victims at Nagasaki/ Hiroshima.

2. Skin cancer occurs more often than any other. Why? The skin is the least important organ of survival and the body can survive the longest when it is compromised. Certainly, any trauma whether physical or

mental can help contribute to toxic load and lead to switch along with a parasitic invasion that signals full-blown cancer.

However, among a select few observers, the sun, when not overdone, now has been shown to have proven anti-cancer effects as a few specialists back away from previously held conclusions that sunlight is dangerous (as a nod to the multimillion dollar sun lotions and blocks industry).

3. Liver cancer is rare. Why? The liver is, by far, the most important organ in the order of survival in the human body as long as the heart keeps pumping. As the above premise states, the entire point behind cancer is an effort to assist in relieving the body of toxins, so the liver is key to this. The last resort then for cancer sites, by this premise, is liver cancer, since heart cancer is completely ruled out as a possibility.

By extension: If you have liver cancer, then you are basically on a final track to death from toxic overload. Therefore, cancer is progressively most deadly according to the where it is located. However, it is always the same "dis-ease." Unlike what we commonly hear then, changing the site does not change the actual condition.

So, cancer is just one condition and the site is related to the toxin, the degree of poison, and where the body intuits it as least damaging or life threatening.

4. The oxidant Chlorine Dioxide, now patented in Europe, http://www.epo.org/searching-for-patents/technical/publication-server.html#tab-1 has been shown to readily kill parasites and stop cancer in its tracks as Royal Rife's work did.

5. MSM, which is one part sulfur, is a strong antitoxin that has been shown to stop cancer as I report below, but this harder to separate as you will see,

Saturated Fats

We are not herbivores. Look at your teeth, please. Meat is a very dense brain food. We need the good fats/ cholesterols in our diets (yes, those saturated fats as shunned for so many years by the mainstream). Getting them back into your diet will help add years to your life as long as they come from clean sources.

Still, there are good reasons for not consuming too many animal fats as follows:

Herein lies the problem today, good clean sources of fats are much harder to come by than even in the '50's. Now that feed lots have become more common and drugs are necessarily employed in the beef producing processes\ just to keep the cattle alive, and fats are the primary storage places for these toxins. However, again note that cats and dogs will fight especially to get uncooked, saturated animal fats as they are dense energy deposits of biological importance just as are the associated organs as determined by their super intelligent quantum bodies (that do not see the toxins).

Continuing with the above, the human body is infinitely more intelligent than our conscious mind (or any doctor's mind). All sub-conscious intelligence, which, as you know, you can't readily affect, is higher knowledge and commonly inaccessible to the reactive lower mind. This is the basis of epigenetics, the newest, yet oldest and most profound of accepted sciences today and QH. So quantum healing as discussed herein, along with the idea of epigentics, is the next step beyond epigenetics. Hang onto your hat as we begin our venture into this next step.

Cancer Cures
Curing Cancer Historically, The Quacks

Why cancer is not cured by doctors and how you might cure it using some historic as well as more recent food and supplement methods:
https://www.youtube.com/watch?v=NAMYAoiCSsI

The Ojibwa Tribe in the above video was likely working with Quantum Science to some degree in their day. Thus, they were the epigeneticists of their time. Essiac tea is simply an herbal mix. However, Essiac tea herbs contain minerals, vitamins, and phytochemicals and these can all deliver the necessary healing nutrients when done correctly, but they also carry with them the vibrations discussed below.

A Proven Case/ An Actual Historic Cure

Essiac does not contain MSM. However, we know that my friend, Paul in New Zealand (possibly the only cured, diagnosed S-4 Mesothelioma Case alive today), used just high levels of (HL) MSM, as described later along with a clean and careful diet.

You also must first realize that, at the time, there was no reason to believe that HL MSM would cure Paul of lung cancer that was apparent. Once you study how MSM works and you understand the Methylation

Pathways (which I did not readily understand then), you know why this protocol works. There is really no mystery here when you study these pathways carefully, but they are indeed challenging.

Furthermore, Paul's cure also should give you some insight into how cancer works as it did me. I suggest that the key is with these Methyl groups, not the sulfur, that really turned the tide for Paul. Still, as Dr. Seneff tells you, the sulfur and cholesterol will help in this. These Methyl groups, when enhanced, raise the body healing vibrations to new levels and thus allow it to perform at the quantum level and heal as introduced in the previous discussion.

Realize also that Paul was first attempting natural dietary cures. Without this combination, no cure would likely have occurred. So, yes, add essiac tea or an herbal mix if you are diagnosed with cancer. Moreover, as reported herein, by using quantum frequencies as previously discussed, instantaneous cancer reversals are possible and even quite likely. What Paul did was to introduce dietary healing frequencies into his body and thus balance his disharmony using this combination as the HL MSM cleared out the offending asbestos as follows:

Paul's problem reportedly began when he breathed in asbestos from brake drums while working on cars as a mechanic over a long period. This above combination, as we now understand it, helped remove this cancer causing toxin from his lungs as it helped restore his vibratory sequences and thus raised his harmonics. While a slow process compared to surgery or radiotherapy, just as with Rich's quantum methods, it is not just a treatment, it is a long-term cure. Thus, unless he were to reintroduce this offending asbestos back into his lungs, Paul is likely cured of this rare form of cancer forever.

Balancing the System with Fats

Finally, sulfur especially, but also cholesterol has ways of binding with toxins and allowing them to pass. Plus the methylation cycles are continually working as energy pathways to clear toxins. The end result, then, is to rebalance the system, raise its harmonic levels, and allow the body the freedom to heal itself, unimpeded, of especially a toxin based problem like cancer.

Historic Cancer Treatments

Some other well known cancer cures and treatments:

The Hoxey Story below is fun, but is only a skin cancer cure. What I like about it is the Morris Fishbine part. Fishbine is the very same guy who ran Royal Rife into bankruptcy. He was the head of the AMA and its spokesman. Were Royal Rife alive today, he could do some of what my friend Rich Price does, but using extremely heavy handed frequency (EMF) methods that he obtained from friends. These were generated by powerful vacuum tube devices, but were effective at killing parasites. He attacked tumors directly and was able to accomplish his cancer cures (still considered incurable by oncologists) in 20 out of 22 cancer cases. Thus, he apparently killed the offending secondary viruses, bacteria, and fungi and freed the patients of biological burden, thus curing their cancer.
https://www.youtube.com/watch?v=90rGPCXWHDk

Dr. Gerson from a totally different angle was absolutely on point. He cured most diseases, including cancer, with just diet, thus was ahead of his time. His book, "A Cancer Treatment," is awesome. The only concern here is with that he limited himself to a vegetarian diet, because he believed that an animal fat diet invites cancer, yet, in fact, few plants possess a full dietary complement. Had he included clean dense food sources like sardines, he may have done better with less effort. Gerson actually cured 50 out of 52 incurable cancer patients.

The AMA hated this man, yet few people are able to even provide the dietary extremes that Gerson used, so he was not a threat to cancer cures. Still, there are other measures that can help equally well or better. Plus, there are new promises in the wind as suggested here that provide more dense food sources such as moringa, cacao, maca, small fish, and any plants raised on our living water that likely are even more ideal.

http://www.amazon.com/Dr-Max-Gerson-Healing-Hopeless-ebook/dp/B002FB66DC

Laetrile may actually work, but will we ever know? One should not be too skeptical, given the real proof. This cure could be explained by the "poison" factor that laetrile introduces, somewhat like mustard gas. Mistletoe (iscatore) injected would fall into the same category.
http://www.chrisbeatcancer.com/b17-laetrile-alternative-cancer-treatment-suppressed-50-years/
Shark cartilage may work also and the theory may prove out, but it is getting long in the tooth and this newer theory below holds much more promise:

The Lipo C idea is best as presented by Dr. Levy below and it can't do a thing to harm as the above poisons could:
https://www.youtube.com/watch?v=z1kD3BolXnE

Next we have another recent promise: "baking soda," as promoted by Dr. Simoncini in Italy. My friend, Dr. Kevin Freeman, MD and I actually talked directly with Simoncini on the phone. Unfortunately, he basically bailed out when he was put to the test. Obviously, cancer has its hangers on who move in when the body's immune system is down, so his argument could hold some water.

We decided that his basic ideas were not far off, but still were too limited to consider as a cure. We know that chemo can keep cancer from ever going away totally, but has always retarded it for a time (why chemo exists) and that some sodium bicarbonate may be in order in getting rid of cancer. Last heard, Simoncini still practices, by the way, by hiring others to do his work.

So the above are all the so called "quacks" who have been disregarded by mainstream medicine. A common mainstream quote is: "We know how cancers behave? (Dr. Barrett, Quackwatch).

So we must ask, "If the above is true, why the dismal record? Dr. Barrett, is a non-practicing psychiatrist making a living off of big pharma (apparently paying him to offend people like Dr. Mercola). Quackwatch (his website) labels them all as quacks. This label is now becoming a badge of honor in the attempt to bring health back to the forefront. Unlike Mercola, Barrett never, apparently, practiced medicine and has been before his own board by reports on the web.

Curing Cancer the Easy Way

Only your body can only heal itself. Cancer is but an expression of toxic overload which limits healthy frequencies. Take away the overload or add back the missing quantum frequency and the body will generally heal. However, despite the popular term, there are no "spontaneous remissions" with cancer. And drugs can never heal the body of any disease as their vibrations are isolated, targeted and they simply distort inherent natural frequencies.

Cancer is but one of the many life threatening diseases. What sets it apart are the costs involved in its treatment and the fact that the few "quacks" have been far more successful at actual cures than all of the mainstream doctors combined.
http://www.cancerdefeated.com/newslette

rs/The-cancer-cure-that-mystifies-doctors.html
http://www.cheniere.org/toc.html

However, given the above Quantum Science, no disease is beyond curing and none can be separated at the quantum level. That is, diseases are all frequency oriented problems. When the missing frequencies are reintroduced, by whatever means, such as replacing the missing essential nutrients or removing the offending toxins, or QH, they cease to exist.

Thus, all diseases are curable as Dan Nelson discusses in his talks. That is, all life threatening diseases go away when the correct vibrational qualities are met. With this new understanding of how vibrations affect us, nothing is impossible when it comes to cures. From the opposite side, no one should become diseased in the first place. All foods are simply vibrations and essential minerals are quite simply, essential vibrations. You can't be in harmony when you are missing essential vibrations, but organic minerals are not the only sources of them. They are simply those most readily available to the general population. They could come from entrained fractals (QH) or other less common source such as transdermal minerals or even getting rid of a bound up spiritual blockage that is hindering vibrations from entering the cells as Dr. Bruce Lipton teaches with epigenetics.

While the above cancer cures may work (or not), herein, I give you a method that will cure any disease, quickly and in a proven way. When you accept this quantum healing (QH) premise, that all disease is frequency based, and you see a disease immediately disappear when these blocks are removed, you simply know. Such results are indisputable. From this, we understand that until now, we have been looking in all of the wrong (most complex) places for cures and no other is needed. Health is always a simplicity!

Common Cold and Flu

Vitamin D is not a vitamin, it is a hormone that the body naturally manufactures from cholesterol and sunlight. This is why the pros warn you not to take too much D3. Cholesterol (and sunlight) scares them. While the toxic levels of D3 have been greatly overstated, one indication that you do not have enough is expressed as a common cold. So if you begin to get cold symptoms, take more. 50,000 IU is easily within the risk factor of most and most can commonly take twice that when a cold is coming on. Doing this commonly results in complete remission and you should never experience another one when you do. I stopped getting even pre-colds many years ago and I take 10k IU/day in winter. Better than D3, sail the BVI winters, get the good fats, and you will never catch colds.

MS (Multiple Sclerosis)

MS is probably one of the most curable of chronic diseases. However, there are at least ten medical conditions commonly mistaken for MS making it one of the most misdiagnosed of diseases. Also, we now have many strains of Lyme running through our population as noted below (people living in the Bronx contract it from pigeon ticks, that mimic MS). MS is not a viral disease, but today we have advanced strains of Lyme that literally defy all common means of diagnosis, thus people are diagnosed with MS. Some forms of Lyme, as noted below, even include viral strains that can never be detected by an MRI or medical diagnostics.

Diagnosis from Beyond the Sub-Conscious Level

While this ability is limited to a few individuals with adequate spiritual training, we predict a time when those "ordained" and trained in these teachings and techniques will become the true healers of the 21st century, a time when invasive disease diagnosis will no longer be accepted or even allowed by society, as awareness of these QH possibilities increase and our drug culture is abandoned.

The above stands in direct opposition to current medicine that makes claims to a paradigm that says, in effect, that medical science can break down nature's herbs and chemicals, then give them a name and patent them, and a doctor can implement them after reading their treatment procedures to cure you of a disease.

The claim here is that this is the best chance any drug has is to hold off the detrimental effects of a disease till the body cures itself. However, this seldom occurs in our real world. In most cases, drugs are never really this successful. The best ones generally mask symptoms till your body does it magic. That is, if it has time, and your vibratory rates are not harmed by the drug itself, you will heal. Drugs mainly serve to relieve pain and symptoms. Any that do more, are exceptional. Since pain is an essential part of the body's defense system in that it warns you of impending problems, masking pain, in itself, creates problems. For instance, if you injure a joint, take a Tylenol, and then walk on it when your body says not to, you injure it worse. Pain exists for a reason and it is not to there to torture you.

Herein, we discuss "Cures," and not pain relief. Our methods are simple. They rely on building autoimmunity through a powerful quantum blend of frequencies that includes natural microorganisms and foods that help you avoid harmful parasites and diseases. This results in a body that literally remains well, is happy both mentally and physically, and thrives in ways never generally seen on this planet for thousands of years except with the very young. However today, even the very young are becoming sickly at a rapid pace.

As this cutting edge science of QH/ Epigenetics catches on and is accepted as scientific fact, as it must, these healings will again become far more common. At first, they will be considered "spontaneous" by the drug culture, because in their minds, a drug or some outside intervention must be used to grossly change a body. While not true, as time goes on, it will become obvious just as it was thousands of years ago and is true with wild animals far from modern influences. More on QH/ Epigenetics below:

Quantum Healing (QH)
My Video on this topic:
https://www.youtube.com/watch?v=qNQrUn7BH3M

On the treatment side, especially when fractals are employed, it is quite mechanical, yet intuitive and scientific in its application and effects. Thus, outcomes are nearly 100% predictable once diagnosed. While these are no more spiritual in nature than healings ever were, healings and anti-aging methods depend on your definition of spirituality, since vibratory frequencies must be included in any definition and seldom are. The premise here is that everyone heals themselves and this can never change.

QH is the ultimate topic here. It incorporates all aspects of this new science of epigenetics (and beyond, as discussed here). Our method of QH includes: fractals, fractals embedded in metals, monatomic metals and an individual program embedded in our lab crystal set (CP4U-F), the very same crystals employed in microchip technology, but employed in tact and unaltered. Within these are induced programs selected to counteract the health concerns and thus raise the vibratory frequencies that relate to the treatment of the condition. Properly done, these produce profound healing effects. Interestingly, these not only include the healing of the condition, in the process, cellular age is lowered, telomeres are lengthened, and microcellular energy factors are raised to levels previously unattainable. Thus, your body is literally driven to healing frequency levels unknown to modern man and you thereby become a superhuman being with a greater energy potential and disease resistance once healed.

The Implications of Quantum Healing

Ancient scripts like the Old Testament Bible speak of the people living to 800 years. The conditions that allowed this to occur are surmised to have included a level of hydration that today's water can not allow.

"Living Water" and "Wayback Water" can. This vibrational level can be temporarily raised by the fractal healing and QH, but to maintain it, one must take use Wayback or Living Water daily. 100% hydration is imperative for most healing to be maintained.

Fractals

Historically, the most simple and commonly known fractal is the star of David, a two-element 2-D geometry comprised of equilateral triangles repeated and inverted. From this, we can expand them to complexities beyond all comprehension as the geometry expands as nature wills it. As a friend suggested as we discussed the Star of

David: Do you think that they knew something in those ancient times? The answer, of course, is: Yes, they knew more than we even envision today for the most part. The real question is: Can we again learn and employ these in our science?

Fractals, the organizing principals of all of nature, can scale up in both their geometry and energy patterns. They can also be enhanced linearly. They represent patterns of sound and light which, spiritually, are the highest, yet most simple, of organizing principals. In fact, fractals serving as filters, are essentially the diagrams of harmonic frequencies (or beats). Thus, they are in no way chaotic as the current Newtonians teach. The reverse is also true. The fractals produced by any healthy biological entity are well organized and their frequencies are harmonically balanced and whole as nature intends. This is the world as it is in all of its beauty. A beautiful face is easily judged as a symmetric fractal and a universal fact.

Thus, if we considered the Star of David as the organizing fractal of a biological entity that is ill, the star might be missing a point and would thus be unbalanced, disorganized, thus chaotic and highly energy consuming (as a leaky dam or roof). Biological entities are, far more complex in their organizing harmonics than this most simple fractal yet both are easily understood intuitively.

The resulting harmonic from a healthy person might be comprised of fifty or more frequencies. If just one frequency is missing or low, the overall harmonic is unbalanced and discordant and generally we thus recognize this person as ill. They consume more energy and are less efficient as their bodies attempt to rebalance. Vibrational imbalance/ rebalance is therefore one correct definition for disease.

Fractals, when employed to heal, are generally balanced geometries that when intuited by a consciousness, filter the harmonics and thus alter the vibrational tones applied. These, then, are the organizing principals that are optimally stacked along with monatomic metals so as to repair and offset the unbalanced frequency. This repairs the hole in the dam that caused the energy losses. Corrective foods and diets in animals, or fuels in machines, also frequencies, can create the same results, but they are slower acting and less dramatic and less certain in their effects than fractals. Dr. Price and I are in a constant search for more powerful fractal combinations for discords.

What else has been forgotten and lost as modern science has trampled on what was once held as sacred fact? These are now considered as subjective religion, and thus disregarded by the mainstream.

Today, we have cherry picked Newtonian Science as the winner and dropped all aspects of the work done by alchemists, all the while complementing ourselves for our brilliance. Alchemists are now commonly considered as misdirected, even somewhat insane, sorcerers, who were said to be attempting to change cheap metals into gold. Misunderstood, it is easy to find fault, as the drug industry now does with naturopaths, especially when it helps their pocketbooks.

So, given the above, verbally all of creation speaks with fractals, but this has been avoided by modern science totally, until just recently. When employed in biological systems that have gone awry, the systems recompose with a new more powerful order. Thus, healings occur instantly and, moreover, health and wellness can be improved to new levels.

Therefore, given this astounding fractal technology, the human body can be improved at the microcellular level to cure all disease, increase energy levels, and generally improve all aspects of life in forgotten ways. Yes, and this is being done today, but the cultural lag is financed by the pharmaceutical and medical industries. We know that they will attempt to hold it back and even kill to maintain their position. Incredibly powerful, nothing in modern times has ever shown the promise that these age-old notions provide, but this is big money. So the real question is, how much longer can these powerful controlling money systems hold back this powerful science? Long term, QH must win, in our opinion. People must eventually get it and no amount of marketing can overcome the facts as they reveal themselves one step at a time.

Interestingly, complex hydraulic, electrical and mechanical systems can be improved to impossible levels when these patterns are applied to them correctly as witnessed on our car tests and in biological systems. When employed, these Quantum systems exceed all of the restrictive laws and limitations that were commonly agreed to and held by today's Newtonian science.

Videos of the CP4U in a car:
https://www.youtube.com/watch?v=YFakNUEQoNM
https://www.youtube.com/watch?v=LcWN_DSmIAI

Our CP4U-F, above increases the energy and efficiency levels of internal combustion engines of all types. Likely, they were the first modern era example of an infinite quantum mechanical/fractal application. This breakthrough device that can be installed on your car in five minutes with no interface or hookup... as a pure quantum device, it has no interface.

In the Egyptian Old Kingdom Era (lost over time) fractals were a commonly accepted method of doing work. In that era, this technology was correctly employed and it accomplished most of the difficult and now considered, impossible, work that was done in making, cutting and moving the massive, nearly perfect, stones in pyramid building.

Pyramids themselves are now commonly misunderstood also. Not tombs, they were necessarily accurate QE energy factories as proven by scientists like Dr. Patrick Flanagan who has written several books on how pyramids actually work and followed by many others since. Quantum devices and fractals demand nearly perfect accuracy to work well. These, of course, have been nearly totally disregarded by the Newtonians also and are forgotten.

Nothing today in what is now called "modern science" nor Newtonian Science can explain how the Egyptians actually moved those stones with such precision or, as with our CP4U-F, how energy output can increase as fuel use decreases and speed increases to 200%+ rates. This CP4U-F technology, when understood, demonstrates that all dynamic engineering systems common today will be outdated once these fractal systems are adapted and we transition into these quantum era applications

Definitions of Fractals that Miss the Mark

Per Google: "Fractals are the infinitely complex patterns that are self-similar across different scales (referred to often as scalar below). They are created by repeating a simple process over and over in an ongoing feedback loop. Driven by recursion, fractals are images of dynamic systems, therein called the pictures of Chaos."

Herein, I offer that Google misses the mark as most of today's science does, joining the ignorant. In fact, fractals only appear chaotic when misunderstood. Scalar energy systems work from the micro to the macro and (thus scale up) generally, just as our CP4U can. Today, these fractal versions are scalar as we implement our programs for specific tasks.

While the web is full of definitions and lectures on fractals by revered physicists (search them from Tom Beardon who reveres scalar energy and can be heard at:

<http://www.cheniere.org/toc.html> to Dr. Bruce Lipton, PhD. Nothing nor anyone out there can actually demonstrate the results reported beyond what Lipton reports at **https://www.brucelipton.com/books/biology-of-belief** from direct experience, but Rich Price and a few others can with QH.

Furthermore, there is very little second party verification other than the survivors of Rich's beneficial healing effects using this fractal technology. Most of them currently have no voice that can be heard above the thundering voice of drugs. So theories abound while applications and results are not currently demonstrated on a large scale. Still, with the arrival of the field of QH/Epigenetics over the last ten years and the welcoming that it has received, we expect a ground swell in acceptance as the cards fall.

Vibratory Frequencies

You are a unique set of higher vibratory frequencies which Dr. Bruce Lipton elegantly describes as beats... like raindrops hitting a pond. These joined result in a single harmonic frequency. This resulting harmonic is sometimes referred to as Soul on the spiritual level. While this may be seen as spiritual in nature, empirically, the frequency levels are measurable and knowable from a quantum viewpoint. So what the ancients were teaching is not only true, it is now actually measurable and scientifically based.

For example, Dan Nelson has found that frequency 28 is always missing when a person is cancerous. For a person to live, frequency 28 must be restored or eventually they will die. Cancer is just one piece of the overall mode of survival that signifies a restricted total harmonic level. So when one raises their overall harmonic level, frequency 28 raises and cancer disappears. Thus, this so called cancerous condition can come and go in seconds. The compromised cellular vibration levels that lead to losing 28 can take years to occur, but your body turns it off and on in order to survive and knows just when to make the required changes to keep you alive as long as possible.

Dis-Ease

As repeated herein in several places, when the various frequencies that comprise an organic (or mechanical) system are out of harmony, it is in a state of Dis-Ease or malfunction. For animals, when these frequencies are compromised, the autoimmune system is also resulting in Dis-

Ease. With Cars that lack Quantum Mechanics (QM), we think of it as normal.

In a human, this disease can be expressed in anything from a cold to life-ending diseases such as cancer, diabetes, and heart disease. These diseases are simply subjective names for missing frequencies, misalignments, and vibratory holes in our cellular systems that are commonly expressed as overall system failures. In plants, slow growth and/or disease results.

All can be cured when the frequencies are brought back into balance or harmony. In QH using a fractal grid they align and this restores the various induced missing vibrational frequencies. Given this, there is no such thing as an incurable disease as Dan Nelson tells us. While an imbalanced condition can be the result of years of dietary abuse or inadequate vibratory conditions, these can all be "cured" in minutes when the body is subjected to the proper balancing frequencies and is thus harmonically rebalanced. This, I have witnessed occur several times recently. Rarely, time reversals can even occur, resulting in totally new outcomes.

Time appears linear and fixed, and to our minds, it is. But time is pliable and reversible under certain conditions. Still, no one decides to reverse time. They occur when you allow them to occur and conditions warrant. Dan Nelson employs a proprietary reverse time laser to make his Wayback water, hence the name.

Cellular Energy Systems/
Mitochondria

You are a complex system of organic microcellular energy systems. The most obvious set is occurring in the mitochondria, the organelles that supply your life energy and allow life to continue. Interestingly, when any animal dies, with these organelles, the overall vibration shuts down in unison. It is just as if the body flipped a light switch. Thus, death is an agreement in every case. Read "Pretty Fine Sex is Spiritual" for a deeper discussion on this topic.

For people (and all biology), when they are not well and have a long term energy problem, the average mitochondrial count (MPC) could be as low as 10, but 100 is a common disease count. Shutting off the above switch with a high energy count is difficult when at 2400. Here you are an unbelievably powerful, disease-free quantum force. At 100, the reverse is true.

Raising the Mitochondrial Count

When the objective is to heal, as discussed, the MCA level correspondingly increases. Thus with the correct fractal set, with the correct intuitive wisdom, it supplies a higher frequency level, causing them to multiply. This is just one form of long-term healing. Even when done properly, raising the MCA to 2400 can take months following treatment. Most people are born with a 700 to 800 count. With this advanced quantum healing, this 2400 level can readily be achieved. But with vital people affected by a parasitic disease, when healed, their MCA can raise to 2400 in days. More on this later.

Bio Subsystems (QB)

If the digestive system is inadequate and lacks the cofactors, the frequencies, even if otherwise available, nothing will never be delivered and your body will simply starve, no matter what your food intake is. Today we know that you are not just a single animal, but a complex biological system that cooperates to keep you in harmony. These cofactors actually supply your cells with B vitamins and other cofactors that you are unlikely to ever obtain otherwise.
When taking antibiotics you kill off this complex life-sustaining system of cofactors that for many years were not even

recognized by modern science. Using antibiotics is just one way of poisoning these necessary biological systems. Radiation events of any kind have similar effects. Thus, MRI or and X-ray testing can set your biology into a downward spiral by killing off these key components and thus short-circuit the required optimum harmonic frequencies (QB) that this biology requires.

Diet

Dietary input is basically frequency input and the term "balanced diet" implies a frequency balance that results in this discussed overall harmonic balance. When you deprive your body of essential frequencies, especially by missing what are termed essential minerals (essential vibrations), it becomes disharmonious. For more on this, listen to Dr. David Perlmutter: **http://www.drperlmutter.com/news/brain change-david-perlmutter/**

As discussed herein and in my previous books, especially, modern farming practices have deprived us of many essential frequencies. Thus, our society is commonly lacking the needed wellness frequencies and is disharmonious, unwell, and diseased. A quick search of the common life-threatening diseases will reveal that they occur more frequently than ever before for this reason.

The Ideal Mix

Thus, with the entire above mix, an optimized quantum healing system will suddenly allow recently unprecedented life spans to occur as long as the necessary frequencies and cofactors are maintained on a daily basis. So how do you do this? Include the food, minerals, and 100% hydration rates, with the vitamins, fats (especially cholesterol as my 1st book and Perimutter, above, points out) and the enzymes required to sustain them and the vibration rates will be sustained. All are simply frequency enhancements and nothing more, just as sugar generally works in opposition to them.

Adding too much of a single factor (over eating), even otherwise ideal foods like nuts or fruits) or eating negative vibrational foods like candy, packaged commercial foods, or sugar loaded drinks (even commercial juices and health drinks) can immediately short-circuit this process. They may not lower your body to a diseased state immediately and it may recover if brought back into balance. However, doing this is additive and eventually this will take its toll, lowering your vibratory rate to that of the norm, which means essentially that you will begin to fail as an organism at about fifty years of age and die at age 75 or thereabouts as dementia sets in and the various systems begin to fail permanently as discussed above.

Telomeres

It has long been recognized that telomeres decrease in length with aging. Also, when scientists measure telomere length, they have determined that biological entities with shorter lengths of telomeres are more disease prone and thus all biological systems are more likely to break down allowing the person to die. Thus, the telomere link is associated with the biological aging clock as noted in the article below.

http://www.alsearsmd.com/ppc/PPC_FOY_proconly.html?utm_source=google&utm_medium=cpc&utm_term=ad&utm_content=buyer&utm_campaign=telomere_foy&gclid=Cj0KEQiAsdCnBRC86PeFkuDJt_MBEiQAUXJfLWTsDCPQocqkGCBJM_VhpGpsY1ePPdiJz90rAGY3Os8aAoGg8P8HAQ

Interestingly, when the quantum healing protocol is employed, telomere length increases substantially along with mitochondria (MCA) as discussed above. Nothing out there can compare to the increases obtained, but there are less effective methods of increasing telomere length and, no, supplementing telomerase will not do what quantum healing does, but it apparently could help. The bottom line is that teleomeres are a function of frequency rates and microcellular harmony, not the reverse, as these scientists propose. However, every aging factor studied thus far, as the above article indicates, has shown that longer telomere length is a direct indicator of health and antiaging, so it may not "cause" health as commonly attributed, but they are certainly a reasonable indicator.

Now, moving along, I introduce recent online discussions involving more conventional issues as they relate to the quantum healing method:

Packaged Supplements/
Heart Disease

Question: "Someone said BBAC (Blockbuster All Clear) is one of the best ways to clear up HBP, but there are others... so can you tell us what they are?"

Read the label. BBAC is a combination of supplements. They are likely well sourced as

it has been shown to help people with clogged arteries and heart disease.
Heart disease (more correctly, circulatory disease), is really just hardened tissue fighting to keep working efficiently, & thus keep you alive. When it can't do that, it fails or parts fail, the chief problem being circulation itself, but also, the circuit becomes leaky from the pressure needed to pump high levels of blood through brittle pipes and valves. The high pressures needed to make the system work eventually take their toll on the system. The body in its wisdom commonly enlarges the heart to compensate, but at some point eventually this fails. This is our most common health problem leading to death with a projected 23.6 million death per year by 2030 if today's "prevention" systems remain in place per the AHA.

The Author takes these interventions regularly, avoiding sugars and grains as reported in , "It's the Liver Stupid." As Dr. Perimutter and other mainstream doctors slowly learn.

I first began studying it in the Yahoo Coconut oil group years ago. Coconut oil, a clean monosaturate, is a good cholesterol builder as discussed above. Dietary changes actually do work, while the AMA and FDA still resist them as they push statins, low cholesterol, and bad fats and even recommend unhealthy foods. The interventions recommended make good sense when you read the chemistry and this is how your body was designed to function.

Also, when taken separately, they are cheaper than BBAC which is posted on their BBAC site with full ingredients. BBAC would be similar to taking a good multivitamin (few exist). As you are a unique biological specimen, if you are really paying attention, your unique blend will surpass any commercial blend such as this one. Also, your combination of supplements will be far better and different than any multivitamin can be, plus any multivitamins are seldom organic or of high quality.

Realize that my Yahoo friend Duncan Crowe was a supplement salesman and he brought this to the table in our discussions and reported on BBAC on his now defunct, but excellent website. However, I suggest that HL MSM will do what BBAC will do (and more), and a low carb diet will simply add to it as will Undenatured Whey. That is, MSM and DMSO, correctly used, will make all connective tissue supple, so that your arteries and veins will become elastic and renew. With this knowledge, there should be no such thing as circulatory disease.

But eating just one vegetable every day is not a good idea and neither is relying on HL MSM alone. Take a full complement. Herbs with their magic phytochemicals synergistically align with the fats such as Coconut oil and Cacao butter. Personally, I take grams of some ten powerful South American herbs each day, organic B's, D3, Undenatured Whey Protein Isolates, Lipo-C, minerals (esp. Mg & Iodine), plus the various components associated with the Methylation Pathways like GABA (Gamma Ammino Butyric Acid), a key contributor. The real key to this is reading and learning about how the pathways work and aiding them. "It's the Liver Stupid 5th edition" is all about this. You must learn the basics.

The above pathways are complex, but universal. That is, while you are an individual in the way that these pathways play out, your microcellular level (quantum level) is virtually universal, except as noted herein. Where we mainly vary is in how we affect them, especially at the subconscious level (epigenetic) and dietary levels. Because of this aspect, when you are exposed to the very same destructive poison as another person, they can raise them to the cancerous level, while you are unaffected. There are many factors involved.

What is claimed above is that cancer is a self protective mode and not really the terrible disease commonly thought. Similarly, by extension, the body uses high blood pressure as a preservation mode. Thus it is raising your blood pressure as a way to keep you alive by combating the hardened tissue and micro-blockages that have occurred as a result of mainly not feeding your body adequate organic sulfur, good fats, and other key ancillary foods. Take blood pressure pills and you fight this life saving action.

However, what you supplement is likely less important than the foods that you avoid... like sugar, potatoes, bread, fried foods, etc. which lower body frequencies. If you play tennis with twenty-year-olds, you have to be very careful with what you take in at age 75. The idea here is to grow wise as you grow old. In doings so, you effectively grow younger and even look younger in this process. Interestingly, as your body grows younger, so does your mental outlook. This is the process that started to prove out ten years ago and what I suggest here as fact. Realize that growing young will create social problems as discussed later here in more detail , but some people are just fine with aging and dying with the population as our government has designed.

Energy Cycles/Balance

Question: "I've tried drinking water with Celtic sea salt, but found it hard going. I watched this video and found a much better way, put a bit on my tongue and then wash it down with lots of water. I love this Aussie. I want to watch all her videos as she knows her science, but she makes it so accessible."
https://www.youtube.com/watch?v=emH O3NekWLw&list=PLX8vMw5EroQIhK VoXKOz9WYFjMGqmhLxL

You may love Barbara O'Neill (and she is indeed very entertaining in her videos). There are a few problems with what she says and she is about where most today are. The biggest problem is with what she leaves out. She is at least ten years behind in her current science (as discussed herein). However, nearly every lecturer on You Tube is also about 10 years behind cutting edge science and she is no exception.

I point out some of the most accurate speakers herein. The most up to date work stems from quantum healing as referenced. If you are ill or bent on staying well, you really must grasp this new level yourself. Barbara is just not going to get you there. Watch Dan Nelson, Richard Gordon (Quantum Healers), and the others that I reference here and elsewhere in my other books. The best that you can normally do is to find one speaker with one up to date conclusion, but Dan has plenty to teach you and Richard generally follows all of my QH conclusions, but mostly spiritually.

Yes, sea water is great stuff, but mainly transdermally, as she suggests in her discussion even though she is not clear on what she has learned herself. But swim in it and please don't drink her combination. The idea of drinking sea water is misconceived even though sea water is full of minerals and micronutrients. Yes, you could absorb seawater in your mouth, but doing it through the skin transdermally is best. Magnesium and organic sulfur are still very important to our health link and both must be supplemented in today's world. I teach swishing HL MSM... an oral absorption method in itself. Seawater could be swished also, but you finally swallow MSM. Skip swallowing seawater often, please.

Think about it: If drinking seawater were a good idea, sailors would never be concerned with drinking it or supplying freshwater on their rafts. They knew that they would die if they did. Barbara's diagram of swelled cells occurs and when seawater is consumed over an extended period, it becomes a poison. Eat sea salts in your food though.

As far as High Blood Pressure (HBP) goes: Barbara is very good up to a point, but she could lead you astray on this also. HBP, as discussed above, is a natural reaction to eating poorly grown food deficient in organic sulfur and saturated fats. This is why HL MSM works so well. It actually softens your arteries and veins and lowers blood pressure as does cholesterol from saturated fat. HBP is a natural adjustment to hardened arteries. It closes off capillaries adding to circulatory problems.

Living Water
Fractally Impregnated Water

Virtually no one in our society is hydrated today. However, to become 100% hydrated, you have two choices: Dan Nelson's water (from his reverse time laser) or Rich's Living Water. Both are comprised of 3-4 nanometer sized particles, but Living Water requires less intake per day to stay hydrated. No other water is close to these when it comes to absorption rates. Four cups of Living Water (less than one liter) per day of this water will keep you 100% hydrated. Rich sells his Living Water pour-thru device plus an Aqua Pure filter for about $600 and you will never have to purchase water or even feel thirsty again in this lifetime. Call him at 302 393 4523, but there is no marketing or fulfillment program in place as of this book edition, so getting your device could take time.

Rich can also sell you a whole-house unit. Then you just turn the tap to get it, plus you can water your garden with it. This water is 400% more powerful (to suit plants), but 400% will easily hydrate people at 100%, still maximizing plant growth. More on that at the end of the book.

So Barbara's intake rates are far too high, but are commonly overstated, since no one generally knows about micro-particle water. Too much water flushes you of electrolytes as she points out, but high intakes tax your kidneys. Electrolytes, which amount to the minerals calcium, sodium, magnesium, potassium, and phosphates dissolved in water are significant keys to your health, especially when you are dehydrated. However, the key is adequate mineral balance and hydration when you are at 100% rates. Less than four cups of low particle water per day will keep you at 100% even when playing tennis in 100deg F weather. I never take water to the courts.

Again, I never carry water to the tennis court and I laugh when watching these people chugging sports drinks (essentially sodas) or drinking quarts of water over just two hour periods. Along the way, I notice that they commonly have elastic bandages on both knees and many half my age are limping. When you use HL MSM, QE, and using Living Water, those days are forgotten. Despite what we see, this is not a fixed issue and, given the proper nutrients, being older is better in many ways if you follow this.

MCA/ Mitochondrial Count Average

Barbara's diagram of cell mitochondria is deficient and here is why: Most of us are not aware of MCA as discussed earlier, why maintaining a high number is so important to energy levels and overall health, or that we can even affect them. Again, you were born, typically, with an 800 MCA and this MCA normally decreases with age. A healthy fifty-year-old will commonly measure 100 MCA. There are naturally higher concentrations in liver cells, because the liver is the key organ and basically oversees all others, while skin cells will have the least MCA where less are needed. However, QH allows us to gain in this critical energy factor and it is easy.

However, we have consistently found that the maximum MCA attainable is about 2400. Also, when cellular frequencies are

Since these mitochondria create the body's energy, your available energy levels increase in direct proportion to mitochondria, while we are naturally told that our energy levels will decrease with the aging process. Commonly low MCA levels sap the body of available energy which we commonly accept as normal today. Thus, the "elderly" lack energy as their skin wrinkles. They loose body mass, and muscles from a lack of nutrient absorption amounting to lower overall vibrational levels. This is not normal, it is sick, we are a sick population. Aging is now spoken of as a sickness and in these terms, it is. It is all interrelated and QH allows you to head it off.

The body's harmonics are balanced by optimal mineral balance or external sources. Can you optimize them just with supplementation? Probably not. Rich uses a "fractal stack" to do this in his healing. Fractals are the geometric QH filters of all the biological harmonics as previously discussed. Your circulatory system is a complex fractal (as are your nervous and lymphatic systems). If one maintains this 2400 MCA level, they can run for miles without being winded at any age, Dan Nelson suggests that he can do this on his video (but he did run into problems with his system). Interestingly, this stuff sticks well.

Barbara is correct in her evaluation of migraines in that dehydration is, apparently, always the cause. If you maintain 96% hydration or more, you will simply never have headaches or migraine attacks. When you become ill you expel electrolytes, so headaches result from the dehydration effects. Migraines are more complex, but hydration is the key to beating them generally.

Barbara's surface tension lecture is mostly correct and much of what she reports on is highly studied by the Scientist Dr. Gerald Pollack and carefully presented in his book: "The Fourth Phase of Water." Realize that Barbara nor Pollack have any knowledge of micro-particle sized water.

Despite Pollack's work and his eminent credentials, reducing surface tension is not the answer to optimum hydration. This was also proposed by another of my other heroes, Dr. Patrick Flanagan. It is all about particle size. That is, Megahydrate, which attempts this, is just not the answer to 100% hydration.

The Dr's Batmangelhedi **https://www.youtube.com/watch?v=oCfDzPs8tvA** and Emoto **https://www.youtube.com/watch?v=Is8FE0RQo8A** did some fascinating work also, but neither ever came up with the real solution when it comes to hydration. As discussed, 100% cellular hydration no longer occurs naturally. The true reason is sourced again from the Quantum Scientist, Dan Nelson here: **https://www.youtube.com/watch?v=7hNW7qxIMzg** So water particle sizes today are simply too large. Until he invented his 3-4 nanometer water 100% hydration was impossible in today's world. We were all, as Dan says, dehydrated. With these two sources, maintaining full hydration is simple. Also, if you are ill, the fact is that nothing will make you fully well till you are 100% hydrated. That is, 100% hydration is just the starting point to vibratory balance. This is where Rich begins when he works his QH on people.

Moving onto Barbara O'Neill's lecture on exercise: **https://www.youtube.com/watch?v=nlxkGzovuEU&index=2&list=PLX8vMw5EroQIhKVoXKOz9WYFjMGqmhLxL&spfreload=10**

Frequency is a vibratory level commonly expressed in hertz (that is, vibratory movements or beats in cycles per second). It does not mean "done frequently" as Barbara implies. This is a cute explanation as it relates to exercise, but this is misleading and the term becomes increasingly important with QH study and understanding.

However, if you are to achieve the levels of health suggested herein, you must not be confused by her statements. In particular: foods, and minerals are simply frequency balancing elements. These elements themselves close the frequency diagram if one were to draw them, just as the Star of David is balanced and closed, but these would appear as a much more complex diagram. Furthermore, when an element is lacking, there is an open fissure or leak in the Methylation Energy Cycle. Thus, the diagram is no longer balanced and disease is commonly the result. Conversely, when complete harmony is achieved, wellness and anti-aging begin, the diagram is closed, and there can be no disease.

Exercise is important, but too much exercise is unbalanced stress, and stress of any kind can set up diseased conditions. The best exercise comes in bursts (abrupt frequency changes), now often termed as "interval training," not just in walking or unstressed exercise, as Barbara suggests. The Mayo clinic describes it at:
http://www.mayoclinic.org/healthy-living/fitness/in-depth/interval-training/art-20044588

Read "Pretty Fine Sex is Spiritual" for a discussion on sex, the very best exercise (smile). Get your spirit right and your body will follow.

Everyone Should Stop Eating These Four Foods!
https://www.youtube.com/watch?v=x5gXRFfMS3g

Dr. Peter Glidden opens the interview discussing the growing number of anti-depressant prescriptions. The discussion leads to the reason for an increased organic mineral intake to prevent disease. This leads Glidden to his "String of Pearls, " discussed below.

This is a good explanation as to why we are a sick nation, mentally and physically, The first person to popularly talk about this was Dr. Joel Wallach of "Dead Doctors Don't Lie," now almost twenty years ago. Minerals (mineral vibrations) are really the bottom line nutritionally, just as he told us then. But now we know that there is plenty more to the wellness and anti-aging formula than just minerals.

Disease Management

This is the String of Pearls that are so easily joined, they tell you why you are not well, even if you think you are. If you are eating the same foods as our sick population, you join the generally sick, even if you are not yet aware of it. We have come to accept this sickness and aging condition as normal and discuss our illness with such terms as, "Its just Growing Pains," or the "Pains of Old Age." No, these are all indeed abnormal. Pain is simply your body screaming, "I am not well!" "Help me please, before I die." "Get me some proper foods with real minerals that I can use and restore the harmonic frequencies that I need to maintain my health."

Peter gives you a fun "gluten" test here that will teach you a lesson quickly if you take it.

Pain reducing drugs are not a solution. When you take them, you are telling your body, "Shut up and just live with what I am feeding you like everyone else." This is what our medical profession, the FDA, and drug companies have degraded us to. Their term: "pain management." A proper diet, free of gluttons and complemented with a full range of organic, bioavailable minerals, will bring love songs to your body and you can throw away your aspirin and all pain killers, which are poisons.

HBP/Optimum Body Weight

The following link is an article from Life Extension Foundation about a high blood pressure (HBP) medicine that has "significantly" lower side effects. But importantly, it has a number of longevity effects. Herein I summarize the important points in the article. However the article does contain a lot of history about HBP medicine and references to trial, studies, etc. Also the article does point out that one should first deal with HBP via diet and life style changes. When that fails to lower the blood pressure, a pharmaceutical drug may be the only solution in their opinion: http:// ww.lef.org/Magazine/2015/3/Best-Drug-To-Treat-Hypertension/Page-01 With this, the question becomes the entire premise that your body does not recognize your correct blood pressure for your inner physical condition (i.e., the pliability and openness of arteries and condition of capillaries and veins). That is, if you have hardened arteries, your body must adjust as discussed previously. In its wisdom, it must maintain a certain blood pressure to maintain the optimum level of health possible for your diseased condition.

Moving your blood pressure down to a typically undiseased condition is just an indication of the mental ignorance of all mainstream treatments. By their premise, lowering your blood pressure artificially is the solution. What you really need to be doing is to soften up your entire circulatory system with a correct diet complement, not by adjusting the symptoms of the hardened dysfunctional system that your body is working feverishly to maintain a life force within.

As to the longevity effects: Body weight falls in line with proper diet. We all recognize that excess body weight indicates ill health. Body fat absorbs D3, arguably one of the most important sources of health. The problem is to know what that weight is, since we are all different with no working manual. However, one only needs to look at animals in the wild to see that none of the herd is ever commonly excessively fat or thin. They adapt or are culled by carnivores. In our case, we just become diseased and fall dead when we deviate too far from our own optimum weight.

EM Devices/ Parasites, Fungi/ Bacteria

A response to a post discussing the Beck electromagnetic healing device not working Comment: "Not noticing any benefits doesn't necessarily mean there is no benefit."

To agree: It would be possible to have bad parasites and be unaware that you have them, so you wouldn't be aware of them being zapped and gone. Having gotten rid of them, you might be spared later problems, but consider why the parasites are there.

It's a similar story with supplements and superfoods, you may not feel any particular benefit, but they may well be keeping your body in balance and preventing disease.

The implication here is that the correct path is to listen to your intuition. There is a part of us that 'knows,' but we tend to ignore that with logic, advice, outside authority, and consensus.

Adding to this: First off, assume that you have parasites, because all people do in various forms, and nearly all people have fungi of some form (closely related).

First, the background: Our preoccupation with parasites took a big hit when the knowledge of beneficial gut bacteria came out in earnest starting in the 1990's. In fact, friendly bacterial are microcellular parasites. They are also significant factors in overall health and make up a large part of your genetic structure. You must have two or more pounds of them to be healthy.

So now we know that we are simply not alone as we assumed since Louis Pasteur passed on his radical antimicrobial theories 150 years ago. Today, it is becoming more clear as time goes on, just how far off base Pasteur was in his conclusions. In fact, we now know that without correct bowel flora, you would be lacking in certain essential vitamins. And as Dr. Perimutter says: Your chances of getting dementia and Alzheimer's goes up considerably.

So, yes, you have parasites, but most are beneficial, while others may be neutral and just stay in balance and cooperate with you without effect. We can now assert with confidence that everyone alive has microbes, fungi, flora and possibly even larger unknown parasites. Still all of these parasites can be cooperating with your body and give you positive effects.

Balance

The key to all of this is that disease prevention is always about balance, not eradication. No wars are needed in the majority of these. We need plenty of anticipation and readiness. It is Oxidation verses Antioxidants and allowing our bodies to seek and play out the magic. But, of course, there are the parasitic diseases like Lyme and Malaria that do attack and kill. The bottom line is then, as always, when healthy balance is lost, the problems begin as with heart disease and cancer the main killers.

When you become ill, other bad guys move in and thrive. This is their areas of expertise since all life attempts to prevail and thrive. These parasites occur with cancer and all major diseases as well as simple dynamic imbalances that occur just from taking drugs and antibiotics.

With the above in mind, you could even overdo Beck's electrification devices. However, these Sota devices, in the hands of a talented, observant healer, can be awesome tools. The key is always in knowing when and where to apply them. In fact, with adequate preparation, it is quite likely that there is no benefit to be gained in killing off bacteria in most cases. So the key is always balance, despite what the brilliant Dr. Robert Beck taught us. For health, a strong defense is far superior to a powerful offense.

Histamines/ Allergies
Natural Solutions
Benedryl

Note that the reactions to (symptoms of) Benedryl are exactly as reported and they include sleepiness, dizziness, drowsiness, confusion, weakness, ringing in the ears, blurred vision, enlarged pupils, dry mouth, flushing, fever, shaking, insomnia, hallucinations, and finally seizures if overdosed enough.

Alcohol combined with Benedryl may increase the drowsiness and dizziness... another reason that you should not drink it in addition to the warnings. We must wonder why people never seem to either read or believe the drug manufacturer's revealed effects. Do they somehow believe that the FDA is protecting them from drugs by allowing drug companies to sell them?

The important thing here is that the common treatment and the foods to avoid in treating allergies and histamines are directly counter to the foods that keep us well and this is a pervasive illness today. So you must make a choice between the good foods for wellness and those that cause allergies and reactions.

Histamine is a biogenetic amine that occurs in various degrees in many foods. In healthy persons, dietary histamine can be rapidly detoxified by amine oxidases, whereas persons with low amine oxidase activity are at risk of histamine toxicity. Diamine oxidase is the main enzyme for the metabolism of ingested histamine. Histamine intolerance occurs when you can't break them down in foods fast enough. It is now known that pickled foods, cheeses, avocados, mushrooms and beets are among the most healthy foods on this planet, so your attention, if you are to remain here for long must be in alleviating this histamine toxicity as soon as possible.

Initially, and as long as you commonly have histamine reactions, it is best to avoid alcoholic beverages altogether, especially beer, champagne and wine (the undistilled forms and commonly the least destructive). Also, generally avoid anchovies, avocados, cheeses, especially aged or fermented cheeses, such as Parmesan, blue and Roquefort, dried fruits such as apricots, dates, figs, prunes, and raisins, fermented foods such as pickled or smoked meats, sauerkraut, mushrooms, processed meats such as sausage, hot dogs, salami, sardines, smoked fish such as herring, pickles, pickled beets, olives, sour cream, sour milk, buttermilk and yogurt (especially if not fresh). So notice, in reading this list, these are the very foods generally recommended for wellness and longevity. This is part of the reason, that histamine and allergy problems are, as mentioned above, a huge issue that short circuits our efforts to stay young even though this is never discussed by mainstream nutritionists, doctors, and especially the drug ads.

However, soured breads, such as pumpernickel, coffee cakes and other foods made with large amounts of yeast, spinach, tomatoes, vinegar or vinegar-containing foods, such as mayonnaise, salad dressing, ketchup, relishes, chili sauce are also on the list and many of these do not make our health list. So all is not bad and there is no doubt that you must be avoiding all forms of bread, cakes, and sugar if you are going to stay young.

We now know that you can naturally acclimate to the above with help from a naturopathic healer who knows this is the new field of epigenetics. Lipo C might help with this (not C), but there are many other natural ways, when you dig down, and some are as follows: Butterbur is a supplement available orally that can help. Mangosteen, a fruit extract supplement, usually sold in juice form or as capsules is also useful (read the label for sugar added). However, Quercetin is one of the best, as the following report by J. Herb Pharmcother suggests in 2003: and quercetin should be on your supplement list.

A water extract of a mixture of eight herbs (chamomile, saffron, anise, fennel, caraway, licorice, cardamom and black seed) was tested for its inhibitory effect on histamine released from rat peritoneal mast cells stimulated either by compound 48/80 or be IgE/anti-IgE. The effect of the herb extract was compared to that of the flavonoid Quercetin. The herbal water-extract inhibited histamine released from chemically and immunologic ally-induced cells by 81% and 85%, respectively; quercetin treated cells were inhibited by 95% and 97%, respectively.

The clinical results showed significant improvements of sleep discomfort, cough frequency, and cough intensity in addition to increased percentages of FEV_1/FVC in patients suffering from allergic asthma, who used the herbal tea compared to those who used the placebo tea, so neither of these are an answer (as we would expect).

Interestingly, coffee, our drink of choice in the US is looking better as a health drink in moderation every day as the reports come in and it may help somewhat here.

As these non-pharmaceutical reports are seldom funded, they are rare.

Still, when you listen carefully to what Dr. Stephanie Seneff tells us about organic sulfur there are adequate clues here if you believe in the HL MSM Protocol to suggest that MSM is the way out of this long term. But hydration is always a key to any of these therapies, so keep that in mind, also.

Next Question: "I get some histamine reactions, such as runny nose (right after eating certain foods). Is that different from a histamine intolerance? What is the origin of histamines? I become sleepy and have low energy and tiredness after certain foods. I tried natural antihistamine supplements: vitamin C and a few others, but sometimes I react to those and also to Benedryl. Yesterday, I had more nausea than usual, so I tried a homeopathic, which helped some. Then, my son noticed my low energy level and suggested eating to raise my blood sugar. That helped me feel better (still nauseous though). I can't always tell that I need to eat."

Allergies are an autoimmune reaction common to a large segment of today's illness accepting society. Over the last hundred years, this segment has grown enormously. One explanation is that, as a society, we are simply too clean and our systems no longer learn to deal with what are now pathogenic bacteria that once were simply dealt with in **balance** as we went through life. So, this is just one of the many ways that our modern methods of dealing with health have backfired badly. In coping with these problems, as noted below, we can actually short-circuit our healthy food choices in an effort to avoid the effects of our lifestyles.

In an allergic reaction, a heterocyclic amine, $C_5 H_9 N_3$, is released by mast cells when tissue is either injured or there is an allergic and inflammatory reaction, causing dilation of small blood vessels. This results in a smooth muscle contraction that can be quite unbearable in many cases as addressed in the question below. These can often be resolved through natural means with some diligence and a little education on your part. But we are not likely to move back to the agricultural lifestyle that most people lived a hundred years ago when they seldom occurred.

I suggest, especially given the recent study below, that high levels of MSM could relieve this over a long enough period and this is one benefit of adequate organic sulfur reintroduced into the diet at the levels that once were fairly common just a hundred years ago. With MSM, the small blood vessels can be made more elastic over time, thus the muscle contractions should begin to disappear completely in due time.

Few currently know this to be true and we are working in a whole new area here, even though it has been reported by many readers who have written me. However, I am fairly certain that this is true for all. Even for those people who have the CBS genetic defect which could initially impede this relief (thus, saying that this will work for most people when employed over a long enough period). Another benefit related to this is that the gut may become more porous and thus increase your hydration rates, thus alleviate the dryness and coughs associated with these conditions. Therefore, 100% cellular hydration can be a huge asset as discussed earlier and almost none maintain that.

MSM Trial /Allergies

While I have referenced sources that claim that allergies are alleviated by MSM. This trial using 2.6 gm/day, minuscule by HL MSM standards, serves to somewhat relieve them. Below I reproduce the report and comment relative to past findings using HL MSM. The point here is that if you have allergies and most everyone in our society today do, it sounds like we have an answer that does not include growing up on the farm and eating dirt like I did:

The MSM Trial

*"In 55 persons with allergic rhinitis (stuffed nose from allergies) given 2600mg MSM supplementation in an open-label trial for one month, allergic symptoms and respiratory complications were reduced by day 7 with symptoms reduction increasing at day 14 (not much more benefit between day 14 and 30) with no alterations in plasma IgE and histamine when subgroups were sampled; the potency was approximately 20-40% symptoms reduction but was not quantified.[42]"***http://examine.com/supplements/Methylsulfonylmethane/**

The point here is that if 2.6 grams per day had some effect, two tablespoons per day or more will be awesome. Given what we have seen when HL MSM which has been used for this and other problems, you can expect an incredibly good outcome. Some readers are reporting total relief from allergies today. I would like to hear more from people with asthma, but I would expect great results with HL MSM over a long period.

Again, allergic or asthmatic conditions are common to many in our society today. While there are other probable factors as noted above, this is also a direct result of our current farming practices that have removed organic sulfur from our soils.

By restoring organic sulfur using HL MSM to the levels that we had before farmers commonly used these destructive fertilizers, these problems can likely be alleviated entirely. It may take you two months to a year or so to build back adequate stores of sulfur, but when you do, you can probably expect complete relief. The HL MSM protocol entails a gradual build-up of organic sulfur till you are taking at least two tablespoons per day if you are over age fifty.

Also, since no heat is used in this delivery system you get all of the sulfur in this MSM and a quicker build-up than nature ever commonly ever provided.

To take it, scoop in two tablespoons full and swish it with water for a long period (I say five minutes), then swallow. Together with alleviating or removing your histamine problem entirely, your general health should improve greatly as you balance your sulfur needs and increase the effectiveness of your methylation cycles. But there are cofactors to add such as: Astaxanthin (hard to get except through supplementation), adequate Magnesium (also hard to get, but cacao is an excellent and tasty natural source), and add

D-3 (adequate sunlight and cholesterol in the body) to the mix.

Watch Dr. Stephanie Seneff's interview on You Tube **http://people.csail.mit.edu/seneff/** Few have a better handle on organic sulfur than Dr. Seneff, but she has no idea about the HL MSM approach in resolving it. She does mention MSM occasionally in her talks saying she that she knows nothing about it.

Seneff gives you a bundle of reasons as to why this MSM is so effective at allergy relief, but one that she does not mention came out in a recent report discussed by Mercola where they discussed that sulfur helped to increase the water absorption in the gut by softening the walls.

Softening tissue is the commanding strength of MSM in all of its applications from arteries to knee tissue to prostate health, so this all makes absolute sense. Certainly, I never took HL MSM hoping to soften my gut tissue, but then I never knew for years that it helps with methylation or cellular energy as it does.

Parasitic Diseases

Malaria... Wikipedia on Malaria:

"In Africa, where 90% of all cases occur, malaria is estimated to result in losses of $12 billion USD a year due to increased healthcare costs, lost ability to work and effects on tourism. The World Health Organization reports there were 198 million cases of malaria worldwide in 2013. This resulted in an estimated 584,000 to 855,000 deaths.
http://www.cdc.gov/malaria/about/facts.html

http://www.who.int/immunization/diseases/malaria/en/

Per the above:
"Malaria is a preventable and treatable (like Lyme) mosquito-borne illness. In 2013, 97 countries had ongoing malaria transmission. There were an estimated 207 million cases of malaria in 2012 (uncertainty range: 135 – 287 million) and an estimated 627,000 deaths (uncertainty range: 473 000 – 789 000). 90% of all malaria deaths occur in sub-Saharan Africa, and 77% occur in children under five. Between 2000 and 2012, an estimated 3.3 million lives were saved as a result of a scale-up of malaria interventions. 90%, or 3 million, of these lives saved are in the under five age group, in sub-Saharan Africa."

There is currently no commercially available malaria vaccine, despite many decades of intense research and development effort. The most advanced vaccine candidate against Plasmodium falciparum is RTS,S/AS01.

A large clinical trial with 15,460 children is ongoing in the following seven countries in sub-Saharan Africa: Burkina Faso, Gabon, Ghana, Kenya, Malawi, Mozambique, and the United Republic of Tanzania." Note: The current falciparum RTS,S/AS01 trial results are indicating that it is 31-56% effective according to age (but for how long?).

Treating Malaria/Actual Case Study

Rich treated a friend using Quantum Fractal Healing and below is the case study which I participated in passively:

The woman treated was 30 years old, but her cellular age prior to the treatment was 48. The Mitochondrial Count Average (MCA) stood at 100 prior to treatment, which is low but common for a person of 48 and especially for a person with malaria. Within four hours after initial treatment, most associated symptoms were gone and by the next day, there were none. Two days after the initial treatment, her cellular age tested at 30 and MCA at just under 2000! Today, it stands at that of an 8 year old and 2400 MCA.

There were no recurring events for three years, but two months ago, she had a reoccurrence and required an additional treatment. The first was during the initial writing of "Beyond Epigenetics."

As parasites are out to save themselves, they mutate and return, so this is common with both Lyme Disease and Malaria. The treatment consisted of placing the woman's picture next to a correctly designed fractal stack and talking to her on the phone. Thus, all is remote. No further episodes were reported till this re-treatment.

Thus MCA corresponded to an energy level of 24 times what it was prior to treatment with a cellular age of 4x less than chronological age. This suggests a permanent cure in most cases, but she did have this reoccurrence. With Malaria, when energy levels are this high and cellular age is low, the body can commonly resist future parasitic infections. This, of course, was not the case, but few people on this planet have ever maintained this energy level.

For those of us who have no experience with a malaria attack her symptoms reported were:
o   Very high fever
o   Not unconscious, but delirious
o   You feel very cold, but sweat profusely
o   These attacks last for maybe a day or so, then subside, unless you die.

Our tests indicate that once a quantum healing has occurred, that cellular aging seldom occurs again, but now we are discussing an infinite biological system and something must occur eventually. Nothing is totally permanent, but these are long lasting.

Both treatments were done using only the person's picture and permission, which is commonly all that Rich needs to do this fractally-based quantum healing. Given that this was his first treatment for this infection, the efficacy rate for malaria is not yet known. However, given that the healing was nearly instantaneous, one could assume that the healing rate could easily stand near 100% for all malarial patients once this QH technique is practiced and taught in Africa.

Having never before treated malaria or these strains of parasites, Rich was not sure that this could even occur, However, I was fairly certain that his Lyme protocols would work for this application, so I ushered him into it and it worked better than we imagined.
The point here was that malaria is not so much different than the various Lyme strains and associated parasites which his techniques have cured routinely. We both learned a great deal from this healing in terms of its associated social implications as noted above from web searches. Malaria is perhaps even more socially disruptive than Lyme, but this is a difficult call since Lyme comes in so many variations and is not so historically recognized as malaria.

My personal interest is in cancer curing, which Rich has done successfully using his quantum healing (along with Alzheimer's to some degree), but we need more cases to make the strong case that this has been made with Lyme and Malaria.

The long term implications of all of this, are, of course, beyond belief. With each healing, we see the same associated anti-aging qualities that accompany the healing, as the MCA is commonly up 24x. My theory is that these techniques kill the associated parasites and the cancer goes away. That is, it eradicates the detrimental flora and the body normally retracts the disease as discussed earlier.

Our current interest will be to train a few thousand other practitioners to use Rich's fractal models and setup according to our QH methods and hopefully, eventually cure all known diseases, especially pathogenic varieties like Lyme and Malaria, which currently evade all known mainstream drugs once they are entrenched (the envisioned dream of genetic science).

Note: no quantum healer will knowingly work with you to heal a person against their will.

MSM Protocol
HL MSM explained/ prototype water device
https://www.youtube.com/watch?v=e8a2m2VO9k0

HL MSM... New Studies and Facts
http://articles.mercola.com/sites/articles/archive/2013/03/03/msm-benefits.aspx

First, with due consideration for Rod Benjamin's marketing program, his more expensive product is not necessary, nor as good as cheaper crystallized MSM brands aimed at animals not denatured by heat. This report reminded me of this one key argument that I had forgotten in my earlier books:

But the first condition was always to use a product that moves quickly. Slower moving, more expensive products pick up more toxic elements in their manufacturing process and especially, storage.

Horse MSM for People

The two MSM brands meant solely for humans employ the more expensive distillation process, as Rod Benjamin proudly reports. Therefore, both by definition, OptiMSM included, are denatured.

Mercola agrees that cleaner sounds like the better option till you understand that heat (denaturing), destroys organic sulfur .

So expense aside, this second reason for avoiding OptiMSM MSM is even more important to my HL protocol. This is, the heat used in the distillation of OptiMSM, Benjamin's company's process likely has detrimental effects on the sulfur just as the farmer who gave me this protocol outlined years ago and warned.

The cheaper crystallization process may be dependant on the manufacturer's water quality as Benjamin says, and, yes, it could contain some contaminants if shortcuts are taken. However, the key here is to find a solution that does not take shortcuts. As it turns out, most all MSM end products actually test very well. But in picking your brand, read their data and follow the first rule above.

Finally, given the levels of use in the HL MSM Protocol, the product price difference is not pennies per dose since we actually use many times his suggested dosage. What Rod suggests or is even aware of is vastly different. That is, OptiMSM at between two tablespoons and ½ cup per day becomes significantly more expensive. Also, as reported in the earlier video, the MSM must be crystalline and "dirty enough to work."

MSM occurs naturally in raw milk, wine, veggies, tea, coffee, etc., Benjamin and Mercola both fail to mention, or do not know, the following: Dr. Stephanie Seneff, (who does not know or did not know about MSM as she admits), clearly points out in her early interviews with Mercola that the

both the heat of cooking and of pasteurization remove organic Sulfur from foods. Listen to her creative thoughts if you have not at:
https://www.youtube.com/watch?v=5QUChSlUEH0 Seneff is a wizard in many areas and her interviews are always very informative, search them all, but start here with this very long seven part series if you have not heard them. These touch on many aspects of health where we have gone badly wrong in the last hundred years.

The bottom line then is to simply buy bulk AniMed or other horse brands, as long as their quality control is maintained as recommended earlier in our HL MSM Protocol. Many seem to miss this point. Notice that Benjamin does mention how raw vegetables contain more sulfur, and this, per Seneff, this is absolutely true. This all speaks against his more expensive product though.

Epigentics/ Diagnosis

First, we now know, through epigentics, that all diseases are mental (sub-conscious) in origin. But we also know through Dr. David Perimuter that doctors have often avoided this fact and that their very diagnosis influences our condition. We have known this for years, but this new science tells us why and also that they can actually affect a cure using their diagnosis. It is all subconscious.

Few doctors today have an objective scientific basis for their dietary conclusions. Our dietary information is based on a long history of incorrect information. Most disease and illnesses are due to negative diet, lifestyle, and the sub-conscious mind... the actual causes of virtually all disease except cancer as discussed previously.

Quantum Healing Aspects
QH/ Socially

Let's say that you are in your mid-seventies, normally within ten years of death and at a time when health breaks downs in today's society. But you test out as age twelve at the cellular level, and your energy levels are through the roof. Your MCA is 2400 and have none of the degenerative physical problems common to your age group (or your spouse's) for obvious reasons.

What do you do? First, you must make one of two choices and there is no other way. All others in your social network who avoid your choice will pass on soon and you will be forced to find new friends and associates and to socially adjust.

Taking the above into account, on an individual level, how should you plan and reorganize your life once you know that you will not be dying anytime soon. Will you find multiple new skills? Should you go back to school and learn a new profession/ broaden your knowledge base? Maybe you combine several skills and professions, knowing that society is bound to change as it picks up on these new possibilities.

When the public commonly supplements the necessary levels of nutrients so that it no longer becomes hypertensive and heart disease is a memory... What then? What will our attention be drawn to as a society when our future outlook is to deal with three hundred or even eight hundred years of wellness?

QH/ Financially

Obviously, the healthcare industry in its present form is going to suffer some mighty blows as the need for managed care evaporates, the drug industry dissolves, and the medical professionals change to incorporate a degree of spirituality for the first time in anyone's knowledge.

This trillion dollar industry, is going to suffer a huge setback as wellness displaces illness as a way of life. Just the fields of oncology at its current $70 billion a year cost and pain management at $60 billion a year will change the face of how both the providers and the patients live.

QH/ Healthcare

When people finally realize that antibiotics are killing their intestinal flora and refuse this method of treatment, then refuse most other drugs as an option... when their bodies are so healthy and immune systems so strong that they no longer normally become infected or even affected with pathogens... Where will this lead?

QH/Government

On a broader level, how will governments adapt as their social services become unnecessary and no longer required? How many years will it take before they realize that the services that they now provide are needless expenses? If our average age suddenly (judging by historic standards), becomes 300 years, how will people cope with wars that wipe out large segments of our population of young people as they have over the last thousand years? Will war be tolerated at all with this new paradigm?

QH/ Legally

How long will it take before our entire legal system, based on life spans of around 70-80 years, adapts to the new paradigm? How will laws change to protect us from accidental death when essentially, this is the only way that people normally die? Finally, What new industries will emerge to encompass our new social needs and expectations?

So these are the problems and questions that we must deal with as we move forward. This new paradigm will and is already beginning to raise these questions for the first time in my own head (in apparently thousands of years) as we enter this new era.

Of course, these social questions will become far more prevalent and profoundly applicable as time goes on and as these new quantum healing principles become more available, recognized, and known. These questions predict that all basic social mores, that is, habits, manners, and attitudes are all about to be questioned (and many times dropped) as new ones displace our long held social values and needs.

QH/ What to Anticipate

So prepare as the current methods of relieving symptoms and pain evaporate. When healing, health, and anti-aging come upon us, be ready. Thus, the word spreads and those around you buy into this new paradigm and benefit. We have only seen only a few hundred healings or so at this point, but as the word spreads prepare to take these questions and outlooks seriously. For now, it may sound like a dream or wish, but it must occur once the money loses its power and the word is spreading among scientists as the following presentations prove:

"Getting to the bottom of Quantum Healing" (the beginning of my next book)

https://video.search.yahoo.com/yhs/search?fr=yhs-pty-pty_email&hsimp=yhs-pty_email&hspart=pty&p=YOu+Tube+Quantum+Mechanics#action=view&id=7&vid=3e0e339f846343e9f8fb32c6b7e32a23

Alan Arkin discusses Quantum Mechanics with four scientists about the two slit experiment and its implications. So here we learn about the dilemma that tells us why these applications have scientific application.

Dr. Bruce Lipton tells us how Epigenetics works

https://video.search.yahoo.com/yhs/search?fr=yhs-adk-adk_sbnt&hsimp=yhs-adk_sbnt&hspart=adk&p=youtube+epigenetics#id=8&vid=49450243edbba32532d7f97939a45db2&action=view

- o Genetics is still an idea and genes do not control our life.
- o Bruce now recognizes that he was teaching dogma in medical school
- o We control our genes. We are not victims and we do not need drugs.
- o The Nucleus (center of DNA) of the cell does not control life.
- o Cells can live for two or more months without genes.
- o Genes are diagrams for life (or blueprints), Genes cause nothing.
- o Medicine is the leading cause of death in the US.
- o Proteins can change their shape... and you can change your proteins epigenetically.
- o The cell membrane controls the proteins {signal transduction}.
- o Cells are miniature humans (our mitochondria run the show really).
- o Signal Transduction.. Half of the DNA is protein and proteins control the DNA strand.
- o DNA does nothing. It is the protein that does the work thru RNA which is a copy system.

- The methyl group (CH3)... how it controls the chromosome by silencing the genes that you do not need (this is the same M as in MSM, folks. More on this of course).
- There is no such thing as junk DNA and this is a key constant.
- Epigenetics can change the signals and cancer is a result of your perceptions. You control everything with your signals (and your subconscious mind).
- You are a skin covered Petri dish. You change the chemistry with your perception and your interpretation.
- The signals are controlled by mind and we are all self biologists.
- How fear is the opposite of love. Fear kills (and I have been writing on this since 1998. It is a given).
- We are the creator of this life!

So on listening to Bruce, we begin to understand how quantum healing (QH) works, but he never uses the term and Bruce never ventures beyond his discovery of the subconscious level in healing. Still his contribution is key to getting quantum healing (QH) out there, which is still considered a religious venture by most scientists still. For some compelling work on this, watch Richard Gordon who is basically there, but Richard has no understanding (or at least recognition) of what these scientists here teach. Put the two together and you have real answers.

As science begins to embrace Epigentics, we now see that everyone is "an Epigenetic Expert." In a talk below, a biochemist and a Nuclear Engineer are both self-proclaimed Epigenetic experts. Notice however that none of these "experts" understand the extent of what Dr. Bruce Lipton teaches, That is, that we can actually do what Moshe Szyf tells us that mothers can do. The key here is that we are in control and not our mothers as Bruce Lipton iterates commonly. Science is moving toward truth slowly, but it is going to be a good while before people like Moshe Szyf surrenders to the facts that Bruce Lipton teaches and Bruce clearly understands what Richard Gordon teaches.

How Epigenetics influences our lives from a biochemist:

https://video.search.yahoo.com/yhs/search?fr=yhs-adk-adk_sbnt&hsimp=yhs-adk_sbnt&hspart=adk&p=youtube+epigenetics#id=9&vid=36bbccd53579e0f25ef0c7c5779c23b4&action=view

- v Biochemical Signals and how they influence our lives.

v   How mothers or surrogates influence the lives of offspring through epigenetics.

Andrew Prentice
https://video.search.yahoo.com/yhs/search?fr=yhs-adk-adk_sbnt&hsimp=yhs-adk_sbnt&hspart=adk&p=andrew+prentice+epigenetics+You+Tube#id=1&vid=87009a319ae91281f2a2eba76a24fcc9&action=click

Andrew is an expert in methylation groups (deep stuff here), but he has no idea about how HL MSM can influence this. Still, we need the cofactors that he discusses that I introduce in my books, The VRNA2-1 Gene must be 100% methylated and without supplementation of HL MSM and the cofactors, these scientists have no knowledge of the supplementation that I teach and it will take them years to get there, if ever. The problem is that they are looking for deep complex solutions rather than something so simple as HL MSM. No digging down is required, no triaging. Sorry Andrew, you will not develop a drug that will replace the simplicity of HL MSM and you will never discover what it is where you are looking.

Randy L. Jertle is an expert in Nuclear Engineering. He has discovered that

Methylation is a key to health and wellness. In this process, he also discovered that low levels of radiation affect the health of mice in a positive way. The interesting thing here is that the questions never go to how methylation affects health, but still he has interesting insight:

https://video.search.yahoo.com/yhs/search?fr=yhs-adk-adk_sbnt&hsimp=yhs-adk_sbnt&hspart=adk&p=youtube+epigenetics#id=11&vid=e6dc597c575da772e10f76049aa72f8e&action=view

Please note that none of the above scientists including Dr, Bruce Lipton, my hero, are aware of the power of high level (HL) MSM even though they all now clearly understand that Methylation is the key factor in the genetic factors and disease susceptibility. Never, today do the so-called experts get that High Level MSM is so outrageously effective because of the Methylation and not just because of the sulfur contribution. Given a choice with an adequate Methyl supply, your intelligent body will make the correct choice as to what to apply.

Next note that Randy Jertle recognizes that water is 80% of the solution in that water interaction encodes the key components of the factors that all of these scientists are looking for. However, this too is not obvious to a scientist who is looking for complex solutions to a simple problem.

Plant Growth and Vibrations (true QB/Quantum Biology)

After proving that cars can be improved dramatically using fractals embedded in lab crystals (QM), then moving this technology to healings, then to increase the hydration levels of water, I noticed that this, now termed "Living Water," was great for plants.

Recently, we found that these fractals in the device itself actually affected the plants just as the water does, since the plants themselves are mostly water and it makes sense that they would. In fact, plants can accept 400% higher energy (vibratory rates) than animals. Rich measures these rates in terms of Life Units (LUs) which he can muscle test (dowses) quickly Watch my video, how to Dowse/ muscle test:
https://www.youtube.com/watch?v=0U7pI14L3sU

Rich is much better than I am at this, but we do test each other occasionally. Both Dan Nelson and Dr. Lipton employ these on their videos. In our experience dowsing is close to 100% accurate for testing existing biological conditions, but only somewhat helpful in predicting outcomes. It is the subconscious mind telling you what is going on. Therefore, the higher your reach spiritually, the more accurate.

My coleus plant has been the test marker for finding all of this. Recently, I placed the Water Device adjacent to the plants and within a day, they migrated toward the device. That is, the plant stems appeared to seek its vibrations over sunlight. The leaves stayed oriented to the sunlight, however. When watered with single pass Living Water as previously, these plants did well, but with both the device and single pass water, you can see by the middle picture, things got better as things improved.

Then on 10/17 I began watering with four pass water (4x the exposure to the embedded fractals and 4000LU). The results of this are in the 11/19 picture. Then, on 11/25 the front plant is flowering over six inch long growths and the second had a one inch flower as it began to flower also.

Check the back cover of this book and seeing the colors and flowers of these coleus plants as they suck up the vibratory energy at rates that no plant has ever had access to in history.

Realize that flowers are a key means of reproduction in plants and when vibratory levels are high, they flower in response to this new energy. While we do not consume coleus plants as food, this occurrence has enormous implications with edible plants and berries that we must sort out.

We previously provided four pass stainless steel devices to high end growers and they initiated these results with one report showing that a cucumber seed produced a cucumber plant with a cucumber in three weeks. Other growers have shown similar results in greenhouses. No one has been given a proximity device to test, but this combination of 400% water and proximity devices could show a new level of productivity previously untested.

These growth rates are a remarkably interesting part of these results. The time to market rates is reduced, thus increasing throughput and crop production rates.

However, the most interesting part of these results is that plants (cucumbers in the first test) grown under these conditions are absorbing 300% higher vibratory rates in their water than any animal can use, making these cucumbers super foods.

This means that when a human eats a cucumber with these high vibratory rates, they can absorb the full rates of the plant, or 3x+ the rates that they can absorb from a Living Water device. This will make these plants incredibly healthy to consume.

Currently, we have no idea what this means in terms of longevity or energy throughput for animals. That is, any plants, berries, or other crops produced by any plants grown under these vibrations could change our lives and those of livestock dramatically. The results of these profound vibrations will most likely be extraordinary. This will take years to sort out logically and scientifically to displace current science, but this is obviously in motion now as the Epigenetic "Experts" learn the whole of the science as Dr. Bruce Lipton has with his work. QB is the new science of our time.

What is Coming?

QB is the new science of our time. We can use it to change everything including adding to our relative time on earth and the quality of the time spent here. We are not a product of our ignorant government, big pharmacy, the medicine it supports, or the terribly inadequate and readily available garbage that is passed on to us as food. Your body knows when you are giving it the right stuff. If it is in pain, you are missing the mark in supplying its needs. Adjust and it will reward you with many comfortable years. But lets review some of the options available and look to the future:

- First, get plenty of MSM and keep the methylation pathways singing.

- Next, when you feel any pain or something is not working as well as it did, rub DMSO into that location. If you were not born with it, your body does not need it. Your body, given the correct nutrients, can reverse any condition. Hearing loss and poor eyesight are not as we are told, a result of aging. They are environmental and you can intervene on your body's behalf. It wants to work well and it will if you help it.

- One of the essential architectural features of all houses that I see coming is the inclusion of a ocean water hot tub/ sunroom with DMSO and other essential nutrients added to the water. Ocean water is an amazing transdermal source of natural minerals. Bathe in hot ocean water with DMSO and you absorb all of the natural minerals that your body craves and in all of the places that you need them. The skin is the delivery system and these are minerals lacking today that other cultures have traveled thousands of miles to gather, but they were general missing some pieces in natural soils so we have always had some disease. With this, you will get the whole nine yards... total wellness. Also with this, we will learn just what is optimal and was never available in the past to the most wealthy people on earth.

- So what else is there after you have the full complement of nutrients? The other half... your ideal spiritual conditions must be met. You cannot be hanging onto the past and, yes, sex is essential along with other forms of exercise and mental cleansing. There is no question that a happy body will reward you with big dividends, but it requires a happy mind and spirit also.

In Closing

We have discussed vibratory rates and healing extensively with this:

I have walked you through the most profound results of this new quantum science and QH likely ever achieved. It is revealing itself everyday with leaps and bounds for disease treatment and has proven itself for machinery (cars), and plants as documented here. In doing this, we have seen that the real answers are indeed "Between the Jeans" and that the Genetic Code does not dictate our health These may be working Jeans but they lack suspenders, they are not working overalls.

Adding to this, as my friend, Dr. Rich Price works with people all over the world and, in many cases, heals them of previously untreatable diseases like Lyme and Malaria (commonly shot at with antibiotics) and with noninvasive methods, often cured in just hours if treated early on.

After proving that cars can be improved dramatically using fractals embedded in lab crystals, then moving this technology to healing, then to increase the hydration levels of water, we now know that Living Water is incredible for plants.

Tell your friends and let's change how things are done just as Dr. Bruce Lipton discusses in his videos.

Videography/ Bibliography:
My videos that document, explain, prove, and complement this book (search: You Tube/ James Robert Clark Quantum Science or cut and paste the following URLs):
CP4U a QM device as used in a car (our first application):
https://www.youtube.com/watch?v=YFakNUEQoNM
https://www.youtube.com/watch?v=LcWN_DSmIAI
Life Extension Results
https://www.youtube.com/watch?v=DAQ_sdcgaBM
Potential Lifespan
https://www.youtube.com/watch?v=eH_wcdu2u-4
https://www.youtube.com/watch?v=06kXgs00Cxw

An Introduction to Spontaneous Evolution" Bruce H. Lipton, PhD

"Back Hole" Nassim Haramein

"Beyond Pyramid Power," Dr. Patrick Flanagan, PhD
ISBN 0-87516-208-8

"The Biology of Belief" Bruce H. Lipton, PhD
https://www.brucelipton.com/books/biology-of-belief

"White Paper" Dan Nelson, PhD
Https://www.waybackwater.com/dans-white-paper

Books below by James Robert Clark addressing health available in Amazon books:
"It's The Liver Stupid," 5th edition
v   Paperback: 324 pages
v   Platform; 5th edition (11/19/17)
v   ISBN-13: 978-1547010493
"Methylation, Awareness and You"
v   Print Length: 114 pages
v   Platform; 2nd edition
v   Publication Date: December 19, 2014
v   BASIN: B00R8P7R3
    "Pretty Fine Sex is Spiritual" (Smile)
    Paperback 182 pages
    Platform 1st Edition
    ASBN-13 978 1495396741

www.ingramcontent.com/pod-product-compliance
Lightning Source LLC
Chambersburg PA
CBHW020441220526
45464CB00002B/800